Real-Time Optimization by Extremum-Seeking Control

Real-Time Optimization by Extremum-Seeking Control

KARTIK B. ARIYUR

MIROSLAV KRSTIĆ

WILEY-INTERSCIENCE

A JOHN WILEY & SONS, INC., PUBLICATION

Library of Congress Cataloging-in-Publication Data:

Ariyur, Kartik B.
    Real-time optimization by extremum-seeking control / Kartic B. Ariyur, Miroslav Kristić.
      p. cm.
    "A Wiley-Interscience publication."
    Includes bibliographical references and index.
    ISBN 0-471-46859-2 (cloth)
    1. Adaptive control systems.   I. Kristić, Miroslav.   II.Title.

TJ217.A665 2003
629.8'36—dc21   2003053460

Printed in the United States of America.

10 9 8 7 6 5 4 3 2 1

Contents

Appendices 199

Preface

Extremum seeking, a popular tool in control applications in the 1940s–1960s, has seen a return as an exciting research topic and industrial *real-time optimization* tool in the 1990's. Extremum seeking is also a method of adaptive control but it does not fit into the classical paradigm or model reference and related schemes, which deal with the problem of *stabilization* of a known reference trajectory or set point.

A second distinction between classical adaptive control and extremum seeking is that the latter is not model based. As such, it provides a rigorous, high performance alternative to control methods involving neural networks. Its non-model based character explains the resurgence in popularity of extremum seeking in the last half a decade: the recent applications in fluid flow, combustion, and biomedical systems are all characterized by complex, unreliable models.

Extremum seeking is applicable in situations where there is a nonlinearity in the control problem, and the nonlinearity has a local minimum or a maximum. The nonlinearity may be in the plant, as a physical nonlinearity, possibly manifesting itself through an equilibrium map, or it may be in the control objective, added to the system through a cost functional of an optimization problem. Hence, one can use extremum seeking both for tuning a set point to achieve an optimal value of the output, or for tuning parameters of a feedback law. The parameter space can be multivariable, a case we cover extensively in this book.

This book overviews the efforts made over the last seven years to put extremum seeking on a rigorous analytical footing and to make improvement of performance in extremum seeking schemes systematic. Stability guidelines that have been developed are applicable not only to static maps but also to systems that combine static maps with dynamics in virtually any form, with the single restriction that the dynamics be open loop stable.[1] The main accomplishment during the recent period, to which this book is dedicated, is achieving convergence to the optimum on a time scale comparable to the time scale of the plant dynamics. In other words, one does not have to try one set of parameters, wait for the plant transient to settle, try another set of parameters, wait again, compare the results, try again, and so on. The convergence of the parameters (set points, gains, etc.) occurs over a period comparable to the length of the plant transients.

Several books in the 1950s–60s have been exclusively or partly dedicated to extremum seeking, including Tsien [109] (1954), Feldbaum [39] (1959), Krasovskii [66] (1963), Wilde [119] (1964), and Chinaev [29] (1969).

A student of history of control should note that extremum seeking was the original method of "adaptive control," first appearing in the 1920s, and developing intensely in the 1940s, especially in the USSR. In the 1960s, extremum seeking branches in two directions. On one side, the emergence of computers (however rudimentary by today's standards) steered the effort on real-time optimization toward general-purpose optimization algorithms. On the other side, a distinction between stabilization and optimization objectives for adaptive control crystallized, and model reference adaptive control methods appeared, which are analytically tractable by simpler, Lyapunov tools. As a result, extremum seeking as a research topic goes dormant for some 30 years. The account of the revival of extremum seeking is given in the Notes and References sections at the ends of chapters.

The book has two parts. Part I is dedicated to comprehensive analysis of perturbation based extremum seeking and its systematic performance improvement. Part II presents various applications that have been successful in recent years and show promise for further development.

Chapter 1 develops stability tests for single parameter extremum seeking and Chapter 2 for multiparameter extremum seeking, and both chapters present systematic design guidelines to satisfy the stability test using standard linear single-input single-output (SISO) control tools. The results render possible rates of adaptation as fast as the plant dynamics. Analysis and design results on extremum seeking are generalized in Chapter 3 to develop slope seeking, a new idea for non-model based control that involves operating a plant at a commanded slope of its reference-to-output map. Chapter 4 presents stability analysis for discrete time extremum seeking with sinusoidal perturbation. Chapter 5 presents the stability analysis for extremum seeking for general non-affine nonlinear systems and Chapter 6 presents stability and performance analysis for limit-cycle minimization via extremum seeking.

The second part of the book presents the successful use of extremum seeking in five applications of immense engineering significance: Chapter 7 presents traction maximization between the wheel and the road; Chapter 8 presents yield maximization of bioreactors; Chapter 9 presents development and application of the first comprehensive design procedure with performance guaran-

---

[1] Open loop unstable dynamics are also admissible by using an internal stabilization loop around which the optimizing, extremum seeking loop, is closed.

tees for minimum power demand formation flight; Chapter 10 presents suppression of gas turbine combustor instabilities in a 4MW combustor at United Technologies Research Center; and Chapters 11 and Chapter 12 present extremum seeking design for maximization of compressor pressure rise, and experimental results on a rig at Caltech along with design for near-maximal compressor pressure rise via slope seeking respectively.

Chapters 5, 6 , 8, 11, and 12 [2] are based on dissertation work of Hsin-Hsiung Wang. Chapter 4 is based on the research of Joon-Young Choi, and Chapter 7 on that of Zhonghua Li. Chapter 10 is joint work of Andrzej Banaszuk and Kartik Ariyur. Chapter 9 is joint work of Paolo Binetti and Kartik Ariyur. The rest of the book is based on the dissertation research of Kartik Ariyur.

Acknowledgments.   In the course of this research, we have benefited from interaction with Andrzej Banaszuk, Paolo Binetti, Joon-Young Choi, Bill Helton, Clas Jacobson, Petar Kokotovic, Richard Murray, Hsin-Hsiung Wang, and Simon Yeung.

We gratefully acknowledge the support that we have received from the Air Force Office of Scientific Research, National Science Foundation, Office of Naval Research, and United Technologies Research Center.

KARTIK B. ARIYUR
MIROSLAV KRSTIĆ

Minneapolis, Minnesota,
and La Jolla, California,
March 2003

---

[2] Excluding Section 12.3.

Part I

THEORY

Chapter 1

SISO Scheme and Linear Analysis

The mainstream methods of adaptive control for linear [7, 46, 57, 78, 89] and nonlinear [69] systems are applicable only for regulation to *known* set points or reference trajectories. In some applications, the reference-to-output map has an *extremum* (i.e., a maximum or a minimum) and the objective is to select the set point to keep the output at the extremum value. The uncertainty in the reference-to-output map makes it necessary to use some sort of adaptation to find the set point which extremizes (maximizes or minimizes) the output. This problem, called *extremum control* or *self-optimizing control*, was popular in the 1950s and 1960s [21, 36, 42, 59, 63, 81, 82, 83, 85, 90, 92], much before the theoretical breakthroughs in adaptive linear control of the 1980's. In fact, the emergence of extremum control dates as far back as the 1922 paper of Leblanc [73], whose scheme may very well have been the first "adaptive" controller reported in the literature. The method of sinusoidal perturbation used in this work has been the most popular of extremum-seeking schemes. In fact, it is the only method that permits fast adaptation, going beyond numerically based methods that need the plant dynamics to settle down before optimization. It is therefore on these extremum seeking schemes that this book is focused.

The purpose of this chapter is to lay a conceptual foundation for extremum seeking and develop familiarity with the methods used for analysis and thereby ease understanding of the more intricate analysis in subsequent chapters. Section 1.1 provides the stability test of the simplest possible extremum seeking scheme—a static map, to ease the reader into the problem. Section 1.2 provides the problem formulation, with Section 1.2.1 supplying linear time-varying (LTV) stability analysis, Section 1.2.2 a linear time-invariant (LTI) stability test, and Section 1.2.3 a design algorithm for a general single parameter extremum seeking scheme. To impress upon the reader the process of design, and the power of the method, we present a simulation study with a difficult

3

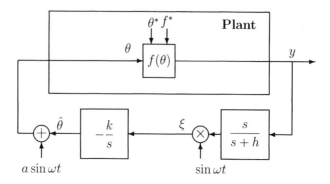

Figure 1.1: Basic extremum seeking scheme

toy model in Section 1.3.

1.1   Extremum Seeking for a Static Map

Figure 1.1 shows a basic extremum seeking loop for a static map. We posit $f(\theta)$ of the form:

$$f(\theta) = f^* + \frac{f''}{2}(\theta - \theta^*)^2 \tag{1.1}$$

where $f'' > 0$. Any C^2 function $f(\theta)$ can be approximated locally by Eqn. (1.1). The assumption $f'' > 0$ is made without loss of generality. If $f'' < 0$, we just replace k $(k > 0)$ in Figure 1.1 with $-k$. The purpose of the algorithm is to make $\theta - \theta^*$ as small as possible, so that the output $f(\theta)$ is driven to its minimum f^*.

The perturbation signal $a \sin \omega t$ fed into the plant helps to get a measure of gradient information of the map $f(\theta)$. We give next an elementary intuitive explanation as to how the scheme "works" with a rigorous analysis to follow in the subsequent sections.

We start by noting that $\hat{\theta}$ in Figure 1.1 denotes the estimate of the unknown optimal input θ^*. Let

$$\tilde{\theta} = \theta^* - \hat{\theta}$$

denote the estimation error. Thus,

$$\theta - \theta^* = a \sin \omega t - \tilde{\theta},$$

which, when substituted into Eqn. (1.1), gives

$$y = f^* + \frac{f''}{2}\left(\tilde{\theta} - a \sin \omega t\right)^2. \tag{1.2}$$

Expanding this expression further, and applying the basic trigonometric identity $2\sin^2 \omega t = 1 - \cos 2\omega t$, one gets

$$y = f^* + \frac{f''}{2}\tilde{\theta}^2 - af''\tilde{\theta}\sin \omega t + \frac{a^2 f''}{2}\sin^2 \omega t \tag{1.3}$$

$$= f^* + \frac{a^2 f''}{4} + \frac{f''}{2}\tilde{\theta}^2 - af''\tilde{\theta}\sin \omega t + \frac{a^2 f''}{4}\cos 2\omega t. \tag{1.4}$$

The washout (high pass) filter

$$\frac{s}{s + h},$$

applied to the output, serves to remove f^*, namely,

$$\frac{s}{s + h}[y] \approx \frac{f''}{2}\tilde{\theta}^2 - af''\tilde{\theta}\sin \omega t + \frac{a^2 f''}{4}\cos 2\omega t. \tag{1.5}$$

This signal is then "demodulated" by multiplication with $\sin \omega t$, giving

$$\xi \approx \frac{f''}{2}\tilde{\theta}^2 \sin \omega t - af''\tilde{\theta}\sin^2 \omega t + \frac{a^2 f''}{4}\cos 2\omega t \sin \omega t. \tag{1.6}$$

As we shall see, it is the second term, and specifically the DC component (or constant) in $\sin^2 \omega t$, that is crucial. Again applying $2\sin^2 \omega t = 1 - \cos 2\omega t$, as well as the identity

$$2\cos 2\omega t \sin \omega t = \sin 3\omega t - \sin \omega t,$$

we arrive at

$$\xi \approx -\frac{af''}{2}\tilde{\theta} + \frac{af''}{2}\tilde{\theta}\cos 2\omega t + \frac{a^2 f''}{8}(\sin \omega t - \sin 3\omega t) + \frac{f''}{2}\tilde{\theta}^2 \sin \omega t. \tag{1.7}$$

Noting that, because θ^* is constant,

$$\dot{\tilde{\theta}} = -\dot{\hat{\theta}},$$

we get

$$\tilde{\theta} \approx \frac{k}{s}\left[-\frac{af''}{2}\tilde{\theta} + \frac{af''}{2}\tilde{\theta}\cos 2\omega t \right.$$
$$\left. + \frac{a^2 f''}{8}(\sin \omega t - \sin 3\omega t) + \frac{f''}{2}\tilde{\theta}^2 \sin \omega t\right]. \tag{1.8}$$

First, we neglect the last term because it is quadratic in $\tilde{\theta}$ and we are interested only in local analysis:

$$\tilde{\theta} \approx \frac{k}{s}\left[-\frac{af''}{2}\tilde{\theta}\right.$$
$$+ \frac{af''}{2}\tilde{\theta}\cos 2\omega t$$
$$\left. + \frac{a^2 f''}{8}(\sin \omega t - \sin 3\omega t)\right]. \tag{1.9}$$

The last two rows are high frequency signals. When passed through an integrator, they get greatly attenuated. Thus, we neglect them, getting

$$\tilde{\theta} \approx \frac{k}{s}\left[-\frac{af''}{2}\tilde{\theta}\right] \tag{1.10}$$

or

$$\dot{\tilde{\theta}} \approx -\frac{kaf''}{2}\tilde{\theta}. \tag{1.11}$$

Since $kf'' > 0$, this is a stable system. Thus, we conclude that $\tilde{\theta} \to 0$, or, in terms of the original problem, $\hat{\theta}(t)$ converges to within a small distance of θ^*.

Looking back at our "hand waving" analysis, it is important to note that our approximations hold only when ω is large (in a qualitative sense) relative to k, a, h, and f''. We shall explore this further in the coming sections. The following bare-bones result sums up the properties of the basic extremum seeking loop in Figure 1.1:

Theorem 1.1 (Extremum Seeking) *For the system in Figure 1.1, the output error $y - f^*$ achieves local exponential convergence to an $O(a^2 + 1/\omega^2)$ neighborhood of the origin provided the perturbation frequency ω is sufficiently large, and $\frac{1}{1+L(s)}$ is asymptotically stable, where*

$$L(s) = \frac{kaf''}{2s}. \tag{1.12}$$

We omit the rigorous proof as this result is subsumed in a more general result we prove in the following section. We draw the reader's attention to the fact that $\frac{1}{1+L(s)}$ is *always* asymptotically stable. We make the wording of this theorem tautological for symmetry with the more general results to come where $\frac{1}{1+L(s)}$ has to be made asymptotically stable by design. The convergence result in the theorem is second order, i.e., $O(a^2 + 1/\omega^2)$, because we are operating around a point of zero slope.

1.2 Single Parameter Extremum Seeking for Plants with Dynamics

Figure 1.2 shows the nonlinear plant with linear dynamics along with the extremum seeking loop. We let $f(\theta)$ be a function of the form:

$$f(\theta) = f^*(t) + \frac{f''}{2}\left(\theta - \theta^*(t)\right)^2, \tag{1.13}$$

where $f'' > 0$ is constant but unknown. The purpose of extremum seeking is to make $\theta - \theta^*(t)$ as small as possible, so that the output $F_o(s)[f(\theta)]$ is driven

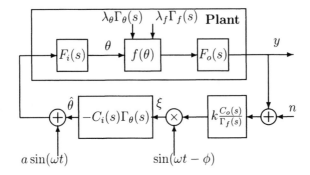

Figure 1.2: Extension of the extremum seeking algorithm to non-step changes in θ^* and f^*

to its extremum $F_o(s)[f^*(t)]$. The optimal input and output, θ^* and f^*, are allowed to be time varying. Let us denote their Laplace transforms by

$$\mathcal{L}\{\theta^*(t)\} = \lambda_\theta \Gamma_\theta(s)$$
$$\mathcal{L}\{f^*(t)\} = \lambda_f \Gamma_f(s).$$

If θ^* and f^* happen to be constant (step functions),

$$\mathcal{L}\{\theta^*(t)\} = \frac{\lambda_\theta}{s}$$
$$\mathcal{L}\{f^*(t)\} = \frac{\lambda_f}{s}.$$

While λ_θ and λ_f are unknown, the Laplace transform (qualitative) form of θ^* and f^* is known, and is included in the washout filter

$$\frac{C_o(s)}{\Gamma_f(s)} = \frac{s}{s+h}$$

(where in the static case we had chosen $C_o(s) = 1/(s+h)$) and in the estimation algorithm

$$C_i(s)\Gamma_\theta(s) = \frac{1}{s}$$

(where in the static case we had chosen $C_i(s) = 1$). Let us first shed more light on $\Gamma_\theta(s)$ and $\Gamma_f(s)$ and then return to discuss $C_i(s)$ and $C_o(s)$.

By allowing $\theta^*(t)$ and $f^*(t)$ to be time varying, we are allowing for the possibility of having to optimize a system whose commanded operation is non-constant. For example, if we have to ramp up the power of a gas turbine engine, we know the shape of, say, $f^*(t)$,

$$\mathcal{L}\{f^*(t)\} = \frac{\lambda_f}{s^2},$$

but we don't know λ_f (and λ_θ). We include $\Gamma_f(s) = 1/s^2$ into the extremum seeking scheme to compensate for the fact that f^* is not constant. The inclusion of $\Gamma_\theta(s)$ and $\Gamma_f(s)$ into the respective blocks in the feedback branch of Figure 1.2 follows the well known *internal model principle*. In its simplest form, this principle guides the use of an integrator in a proportional-integral (PI) controller to achieve a zero steady-state error. When applied, in a very generalized manner, to the extremum seeking problem, it allows us to track time-varying maxima or minima.

We return now to the compensators $C_o(s)$ and $C_i(s)$. Their presence is motivated by the dynamics $F_o(s)$ and $F_i(s)$, but also by the reference signals $\Gamma_\theta(s)$ and $\Gamma_f(s)$. For example, if we are tracking an input ramp,

$$\Gamma_\theta(s) = \frac{1}{s^2},$$

we get a double integrator in the feedback loop, which poses a threat to stability. Rather than choosing $C_i(s) = 1$, we would choose $C_i(s) = s + 1$ (or something similar) to reduce the relative degree of the loop. The compensators $C_i(s)$ and $C_o(s)$ are crucial design tools for satisfying stability conditions and achieving desired convergence rates.

We now make assumptions upon the system in Figure 1.2 that underlie the analysis to follow:

Assumption 1.2 $F_i(s)$ and $F_o(s)$ are asymptotically stable and proper.

Assumption 1.3 $\Gamma_f(s)$ and $\Gamma_\theta(s)$ are strictly proper rational functions and poles of $\Gamma_\theta(s)$ that are not asymptotically stable are not zeros of $F_i(s)$.

This assumption forbids delta function variations in the map parameters and also the situation where tracking of the extremum is not possible.

Assumption 1.4 $\frac{C_o(s)}{\Gamma_f(s)}$ and $C_i(s)\Gamma_\theta(s)$ are proper.

This assumption ensures that the filters $\frac{C_o(s)}{\Gamma_f(s)}$ and $C_i(s)\Gamma_\theta(s)$ in Figure 1.2 can be implemented. Since $C_i(s)$ and $C_o(s)$ are at our disposal to design, we can always satisfy this assumption. The analysis does not explicitly place conditions upon the dynamics of the parameters $\Gamma_\theta(s)$ and $\Gamma_f(s)$, however, for any design to yield a nontrivial region of attraction around the extremum, they cannot be faster than plant dynamics in $F_i(s)$ and $F_o(s)$. The signal n in Figure 1.2 denotes the measurement noise.

1.2.1   Single Parameter LTV Stability Test

We first provide background for the result on output extremization below. The equations describing the single parameter extremum seeking scheme in

Figure 1.2 are:

$$y = F_o(s)\left[f^*(t) + \frac{f''}{2}(\theta - \theta^*(t))^2\right] \tag{1.14}$$

$$\theta = F_i(s)\left[a\sin(\omega t) - C_i(s)\Gamma_\theta(s)[\xi]\right] \tag{1.15}$$

$$\xi = k\sin(\omega t - \phi)\frac{C_o(s)}{\Gamma_f(s)}[y + n]. \tag{1.16}$$

For the purpose of analysis, we define the tracking error $\tilde{\theta}$ and output error $\tilde{y}$:

$$\tilde{\theta} = \theta^*(t) - \theta + \theta_0 \tag{1.17}$$

$$\theta_0 = F_i(s)\left[a\sin(\omega t)\right] \tag{1.18}$$

$$\tilde{y} = y - F_o(s)[f^*(t)]. \tag{1.19}$$

In terms of these definitions, we can restate the goal of extremum seeking as driving output error $\tilde{y}$ to a small value by tracking $\theta^*(t)$ with θ. With the present method, we cannot drive $\tilde{y}$ to zero because of the sinusoidal perturbation θ_0. We now provide a minimally restrictive stability test for the system in Figure 1.2:

Proposition 1.5 (Single Parameter Extremum Seeking: LTV Test)
For the system in Figure 1.2, under Assumptions 1.2, 1.3, and 1.4, the output error $\tilde{y}$ achieves local exponential convergence to an $O(a^2)$ neighborhood of the origin provided $n = 0$ and:

1. *$\pm j\omega$ is not a zero of $F_i(s)$.*

2. *Zeros of $\Gamma_f(s)$ that are not asymptotically stable are also zeros of $C_o(s)$.*

3. *Poles of $\Gamma_\theta(s)$ that are not asymptotically stable are not zeros of $C_i(s)$.*

4. *$C_o(s)$ is asymptotically stable and the eigenvalues of the matrix $\mathbf{\Phi}(T,0)$ lie inside the unit circle, where $T = 2\pi/\omega$ and $\mathbf{\Phi}(T,0)$ is the solution at time T of the matrix differential equation*

$$\dot{\mathbf{\Phi}} = \mathbf{A}(t)\mathbf{\Phi}(t,0), \quad \mathbf{\Phi}(0,0) = \mathbf{I}, \tag{1.20}$$

and $\dot{\mathbf{x}} = \mathbf{A}(t)\mathbf{x}(t,0)$, $\mathbf{x}(0) = \mathbf{x}_0$ is a state space representation of the LTV differential equation

$$\mathrm{den}\{H_i(s)\}[\tilde{\theta}] = -f''\mathrm{num}\{H_i(s)\}\left[\sin(\omega t - \phi)H_o(s)\left[\theta_0(t)\tilde{\theta}\right]\right], \tag{1.21}$$

where

$$H_i(s) = C_i(s)\Gamma_\theta(s)F_i(s)$$

$$H_o(s) = k\frac{C_o(s)}{\Gamma_f(s)}F_o(s).$$

Proof of Proposition 1.5: Setting $n = 0$ and substituting Eqns. (1.15) and (1.18) in Eqn. (1.17) yields

$$\tilde{\theta} = \theta^* + H_i(s)[\xi] \tag{1.22}$$

Further, substitution for ξ from Eqn. (1.16) and for y from Eqn. (1.14) yields

$$\tilde{\theta} = \theta^* + H_i(s)\left[\sin(\omega t - \phi)H_o(s)\left[f^* + \frac{f''}{2}(\theta - \theta^*)^2\right]\right]. \tag{1.23}$$

Using $\theta - \theta^* = \theta_0 - \tilde{\theta}$ from Eqn. (1.17), we get

$$
\begin{aligned}
\tilde{\theta} &= \theta^* + H_i(s)\left[\sin(\omega t - \phi)H_o(s)\left[f^* + \frac{f''}{2}(\theta_0 - \tilde{\theta})^2\right]\right] \\
&= \theta^* + H_i(s)\left[\sin(\omega t - \phi)H_o(s)\left[f^* + \frac{f''}{2}(\theta_0^2 - 2\theta_0\tilde{\theta} + \tilde{\theta}^2)\right]\right]. \tag{1.24}
\end{aligned}
$$

We drop the higher order term $\tilde{\theta}^2$ (this is justified by Lyapunov's first method, as we have already written the system in terms of error variable $\tilde{\theta}$ thus transforming the problem to stability of the origin) and simplify the expression in Eqn. (1.24) using Lemmas A.1, A.2, Assns. 1.2, 1.3, and 1.4 and asymptotic stability of $\frac{C_o(s)}{\Gamma_f(s)}$ and $C_o(s)$:

$$
\begin{aligned}
\sin(\omega t - \phi)H_o(s)[f^*(t)] &= \lambda_f \sin(\omega t - \phi)\mathcal{L}^{-1}\left(H_o(s)\Gamma_f(s)\right) \\
&= \sin(\omega t - \phi)(\epsilon^{-t}) = \epsilon^{-t} \tag{1.25}
\end{aligned}
$$
$$\sin(\omega t - \phi)H_o(s)[\theta_0^2] = C_1 a^2 \sin(\omega t + \mu_1) + C_2 a^2 \sin(3\omega t + \mu_2) + \epsilon^{-t} \tag{1.26}$$

where C_1, C_2, μ_1, μ_2 are constants (these can be determined from the frequency response of $H_o(s)$), and ϵ^{-t} denotes exponentially decaying terms. Now denote

$$u_{13}(t) = a^2\frac{f''}{2}\left[C_1 \sin(\omega t + \mu_1) + C_2 \sin(3\omega t + \mu_2)\right]. \tag{1.27}$$

The tracking error equation, Eqn. (1.24) after linearization (effected simply by dropping $\tilde{\theta}^2$ terms as we have expressed the system as an ODE in $\tilde{\theta}$) can be rewritten as

$$\tilde{\theta} = \theta^* + H_i(s)\left[u_{13}(t) + \epsilon^{-t} - f''\sin(\omega t - \phi)H_o(s)\left[\theta_0\tilde{\theta}\right]\right]. \tag{1.28}$$

Multiplying both sides of Eqn. (1.28) with the denominator of $H_i(s)$, we get

$$\text{den}\{H_i(s)\}[\tilde{\theta}] = \epsilon^{-t} + \text{num}\{H_i(s)\}\left[u_{13}(t) + \epsilon^{-t} - f''\sin(\omega t - \phi)H_o(s)\left[\theta_0\tilde{\theta}\right]\right]. \tag{1.29}$$

The θ^* term drops out or becomes an exponentially decaying term when operated upon by den$\{\Gamma_\theta(s)\}$ contained in den$\{H_i(s)\}$. We now write a state space representation of the LTV system in Eqn. (1.29):

$$\dot{\mathbf{x}} = \mathbf{A}(t)\mathbf{x} + \mathbf{B}u_{13}(t), \quad \mathbf{A}(t+T) = \mathbf{A}(t), \quad T = 2\pi/\omega. \qquad (1.30)$$

The system has a state transition matrix $\boldsymbol{\Phi}(t,0)$ given by the solution of

$$\dot{\boldsymbol{\Phi}} = \mathbf{A}(t)\boldsymbol{\Phi}(t,0), \quad \boldsymbol{\Phi}(0,0) = \mathbf{I}. \qquad (1.31)$$

The system is exponentially stable if the eigenvalues of the matrix $\boldsymbol{\Phi}(T,0)$ (numerically calculated above) lie within the unit circle by Property 5.11 in reference [97]. As the persistent part of the non-homogenous forcing term in Eqn. (1.29) is $O(a^2)$, we have convergence of $\tilde{\theta}$ to $O(a^2)^1$ and therefore the convergence of $\tilde{y} = y - F_o(s)[f^*(t)] = F_o(s)\left[f''/2(\tilde{\theta} - \theta_0)^2\right]$ to $O(a^2)$. Q.E.D.

1.2.2 Single Parameter LTI Stability Test

While the result above permits determination of stability of extremum seeking loops in a wide variety of cases, it is not a convenient *design* tool as it requires calculation of the state transition matrix of an LTV system. For this purpose, we provide a result below that permits systematic design in a variety of situations. To this end, we introduce the following notation:

$$H_o(s) = k\frac{C_o(s)}{\Gamma_f(s)}F_o(s) \triangleq H_{osp}(s)H_{obp}(s) \triangleq H_{osp}(s)(1 + H_{obp}^{sp}(s)) \ (1.32)$$

where $H_{osp}(s)$ denotes the strictly proper part of $H_o(s)$ and $H_{obp}(s)$ its biproper part, and k is chosen to ensure

$$\lim_{s \to 0} H_{osp}(s) = 1. \qquad (1.33)$$

Now we make an additional assumption upon the plant:

Assumption 1.6 *Let the smallest in absolute value among the real parts of all of the poles of $H_{osp}(s)$ be denoted by a. Let the largest among the moduli of all of the poles of $F_i(s)$ and $H_{obp}(s)$ be denoted by b. The ratio $M = a/b$ is sufficiently large.*

---

[1]Exponential stability of the homogenous part of the linear system in Eqn. 1.29 implies $\mathcal{L}$-stability of the system with forcing. Boundedness of the inhomogenous forcing yields boundedness of the solution of the forced system as in Example 6.3 in Khalil [64], and the bound on the forcing gives the order of $\tilde{\theta}$.

With this assumption, we separate the slow and fast dynamics in the LTV system in Eqn. (1.29) as

$$\text{den}\{H_i(s)\}[\tilde{\theta}] = \epsilon^{-t} + \text{num}\{H_i(s)\} \left[u_{13}(t) + \epsilon^{-t} - f'' \sin(\omega t - \phi))y'_{osp}\right]$$

$$y'_{osp} = (1 + H_{obp}^{sp}(s))[y_{osp}] \tag{1.34}$$

$$y_{osp} = H_{osp}(s) \left[\theta_0 \tilde{\theta}\right], \tag{1.35}$$

and write the following state-space representation for the fast dynamics:

$$\frac{1}{M}\dot{\mathbf{x}}_{osp} = \mathbf{A}_{osp}\mathbf{x}_{osp} + \mathbf{B}_{osp}[\theta_0\tilde{\theta}] \tag{1.36}$$

$$y_{osp} = \mathbf{C}_{osp}\mathbf{x}_{osp}, \tag{1.37}$$

where the eigenvalues of $M\mathbf{A}_{osp}$ are the poles of $H_{osp}(s)$. Reduction of the fast dynamics in Eqn. (1.36) by singular perturbation yields the substitution

$$y_{osp} = \mathbf{C}_{osp}\mathbf{A}_{osp}^{-1}\mathbf{B}_{osp}[\theta_0\tilde{\theta}] = \theta_0\tilde{\theta}, \tag{1.38}$$

using Eqn. (1.33) to form a reduced order model.

The purpose of this assumption is to use a singular perturbation reduction of the output dynamics and provide the LTI SISO stability test of the theorem stated below. If the assumption were made upon the output dynamics $F_o(s)$ alone, the design would be restricted to plants with fast output dynamics $F_o(s)$. Hence, for generality in the design procedure, the assumption of fast poles is made upon the strictly proper factor $H_{osp}(s)$ of $H_o(s)$. Its purpose is to deal with the strictly proper part of $F_o(s)$. If we have slow poles in a strictly proper $F_o(s)$, *we can introduce a biproper* $\frac{C_o(s)}{\Gamma_f(s)}$ *with an equal number of fast poles* to permit analysis based design. For example, if

$$F_i(s) = \frac{1}{s+1}, \text{ and } F_o(s) = \frac{1}{(s+1)(2s+3)},$$

with constant f^* and θ^* (giving $\Gamma_\theta(s) = \Gamma_f(s) = 1/s$) we may set

$$C_o(s) = \frac{(s+4)}{(s+5)(s+6)}$$

and $k = 60$ to give

$$H_o(s) = \frac{C_o(s)}{\Gamma_f(s)}F_o(s) = \frac{60s(s+4)}{(s+1)(2s+3)(s+5)(s+6)}.$$

We can factor the fast dynamics as

$$H_{osp}(s) = \frac{30}{(s+5)(s+6)}$$

and the slow biproper dynamics as

$$H_{obp}(s) = 1 + H_{obp}^{sp}(s) = 1 + \frac{1.5(s-1)}{(s+1)(s+1.5)}.$$

This gives, in the terms of Assumption 1.6, the smallest pole in absolute value in $H_{osp}(s)$, $a = 5$, the largest of the moduli of poles in $F_i(s)$ and $H_{obp}(s)$ as $b = 1.5$, giving their ratio $M = a/b = 3.33$. The singular perturbation reduction reduces the fast dynamics $H_{osp}(s) = \frac{30}{(s+5)(s+6)}$ to its unity gain, and we deal with stability of the reduced order model via the method of averaging to deduce stability conditions for the overall system in the theorem below. Hence, for all situations, we can first perform systematic design, and then, if necessary, reduce the order of the output filter and check for stability using the LTV test above.

To keep the proof brief, we make an additional assumption:

Assumption 1.7 $H_i(s)$ *is strictly proper.*

This assumption is very easy to satisfy. Either $F_i(s)$ is strictly proper or, if it is biproper, one would choose $C_i(s)\Gamma_\theta(s)$ strictly proper. For example, if $F_i(s)$ is biproper and $\Gamma_\theta(s) = 1/s$, $C_i(s) = 1$ satisfies this assumption. The assumption is made only for the purpose of keeping the proof of the following theorem brief. The formation of a state space representation of the reduced order system for averaging becomes more intricate when $H_i(s)$ is biproper, because of the need to account for a factor of ω when time varying terms are differentiated, and this distracts from the main theme of the proof.

Theorem 1.8 (Single Parameter Extremum Seeking) *For the system in Figure 1.2, under Assumptions 1.2–1.7, the output error $\tilde{y}$ achieves local exponential convergence to an $O(a^2 + \delta^2)$ neighborhood of the origin, where $\delta = 1/\omega + 1/M$ provided $n = 0$ and:*

1. *Perturbation frequency ω is sufficiently large, and $\pm j\omega$ is not a zero of $F_i(s)$.*

2. *Zeros of $\Gamma_f(s)$ that are not asymptotically stable are also zeros of $C_o(s)$.*

3. *Poles of $\Gamma_\theta(s)$ that are not asymptotically stable are not zeros of $C_i(s)$.*

4. *$C_o(s)$ and $\frac{1}{1+L(s)}$ are asymptotically stable, where*

$$L(s) = \frac{af''}{4}\mathrm{Re}\{e^{j\phi}F_i(j\omega)\}H_i(s). \qquad (1.39)$$

Proof. We rewrite the linearized model Eqn. (1.28)[2] after reduction of the fast dynamics $H_{osp}(s)$ to its unity static gain (from Assumption 1.6 and Eqn. (1.38)) and get

$$\tilde{\theta} = \theta^* + H_i(s) \left[u_{13}(t) + \epsilon^{-t} - f'' \sin(\omega t - \phi)(1 + H_{obp}^{sp}(s)) \left[\theta_0 \tilde{\theta} \right] \right]. \quad (1.40)$$

Using Lemmas A.1, and A.2, we obtain[3]

$$\tilde{\theta} = \theta^* + H_i(s) \left[u_{13}(t) + \epsilon^{-t} - af''/4 \operatorname{Re}\{e^{j\phi} F_i(j\omega)\}\tilde{\theta} + v_1 + v_2 \right] \quad (1.41)$$

$$v_1 = af''/4 \operatorname{Re}\{e^{j(2\omega t - \phi)} F_i(j\omega)\}\tilde{\theta} \quad (1.42)$$

$$v_2 = \sin(\omega t - \phi) H_{obp}^{sp}(s) [v_3] \quad (1.43)$$

$$v_3 = |F_i(j\omega)| \sin(\omega t + \angle F_i(j\omega))\tilde{\theta}. \quad (1.44)$$

We next move the term $af''/4 \operatorname{Re}\{e^{j\phi} F_i(j\omega)\} H_i(s)\tilde{\theta} = L(s)\tilde{\theta}$ to the left hand side, and divide both sides of Eqn. (1.41) with $1 + L(s)$:

$$\begin{aligned} \tilde{\theta} &= \frac{1}{1+L(s)}[\theta^*] + Y_i(s) \left[u_{13}(t) + \epsilon^{-t} + v_1 + v_2 \right] \\ &= \epsilon^{-t} + Y_i(s) \left[u_{13}(t) + v_1 + v_2 \right], \end{aligned} \quad (1.45)$$

where $Y_i(s) = \frac{H_i(s)}{1+L(s)} = \frac{\operatorname{num}\{Y_i(s)\}}{\operatorname{num}\{1+L(s)\}}$ is asymptotically stable because the poles of $H_i(s)$ are cancelled by zeros of $\frac{1}{1+L(s)}$, and $\frac{1}{1+L(s)}$ is asymptotically stable; zeros in $\frac{1}{1+L(s)}$ cancel poles in $\theta^*(s) = \lambda_\theta \Gamma_\theta(s)$ resulting in exponentially decaying terms which are all consolidated (the asymptotically stable transfer function $Y_i(s)$ acting on exponentially decaying terms produces exponentially decaying terms). Now, $Y_i(s)$ is strictly proper from Assumption 1.7, and can therefore be written as $Y_i(s) = \frac{1}{s+p_0}Y_i'(s)$, where $Y_i'(s)$ is proper and p_0 is a pole of $Y_i(s)$ (and therefore asymptotically stable). In terms of their partial fraction expansions, we can write $Y_i'(s) = A_0 + \sum_{k=1}^{n} \frac{A_k}{s+p_k}$, and $H_{obp}^{sp}(s) = \sum_{j=1}^{m} \frac{B_j}{s+p_j}$. Multiplying both sides of Eqn. (1.45) with $s+p_0$ and using the partial fraction expansions, we get

$$\dot{\tilde{\theta}} + p_0 \tilde{\theta} = \epsilon^{-t} + A_0(u_{13}(t) + v_1 + v_2)$$

---

[2]The use of Lyapunov's first method in the proof of Proposition 1.5 and here is what makes the stability result of Theorem 1.8 local.

[3]Note that Eqn. (1.41) contains an additional term of the form $H_i(s)[\sin(\omega t - \phi)H_o(s)[\epsilon^{-t}\tilde{\theta}]]$ which comes from ϵ^{-t} in $\theta_0(t) = a\operatorname{Im}\{F_i(j\omega)e^{j\omega t}\} + \epsilon^{-t}$. We drop this term from subsequent analysis because it does not affect closed loop stability or asymptotic performance. It can be accounted for in three ways. One is to perform averaging over an infinite time interval in which all exponentially decaying terms disappear. The second way is to treat $\epsilon^{-t}\tilde{\theta}$ as a vanishing perturbation via Corollary 5.4 in Khalil [64], observing that ϵ^{-t} is integrable. The third way is to express ϵ^{-t} in state space format and let $\epsilon^{-t}\tilde{\theta}$ be dominated by other terms in a local Lyapunov analysis.

$$+ \sum_{k=1}^{n} [u_{13,k}(t) + v_{1,k} + v_{2,k}] \tag{1.46}$$

$$v_2 = \sin(\omega t - \phi) \sum_{j=1}^{m} (v_{3,j})$$

$$u_{13,k} = \frac{A_k}{s + p_k} u_{13}, \quad v_{1,k} = \frac{A_k}{s + p_k} v_1$$

$$v_{2,k} = \frac{A_k}{s + p_k} v_2, \quad v_{3,j} = \frac{B_j}{s + p_j} v_3.$$

Now, we can write the system above in state-space form[4]:

$$\dot{\mathbf{x}} = A(t)\mathbf{x} + A_{12}\mathbf{x}_e + Bu_{13}(t); \quad \tilde{\theta} = x_1 \tag{1.47}$$

$$\dot{\mathbf{x}}_e = A_e\mathbf{x}_e. \tag{1.48}$$

Eqn. (1.48) is a representation for the ϵ^{-t}. We get Eqns. (1.47), (1.48) into the standard form for averaging by using the transformation $\tau = \omega t$, and then averaging the right hand side of the equations w.r.t. time from 0 to $T = 2\pi/\omega$, i.e., $\frac{1}{T} \int_0^T (\cdot) d\tau$ treating states $\mathbf{x}$, $\mathbf{x}_e$ as constant to get:

$$\frac{d\mathbf{x}_{av}}{d\tau} = \frac{1}{\omega} (A_{av}\mathbf{x}_{av} + A_{12}\mathbf{x}_{eav}), \quad \tilde{\theta}_{av} = x_{1av} \tag{1.49}$$

$$\frac{d\mathbf{x}_{eav}}{d\tau} = \frac{1}{\omega} A_e\mathbf{x}_{eav}, \tag{1.50}$$

which is a state-space representation of the system in the $\tau = \omega t$ time-scale, and $A_{av} = \frac{1}{T} \int_0^T (A(\tau)) d\tau$. This gives

$$\dot{\tilde{\theta}}_{av} + p_0\tilde{\theta}_{av} = \epsilon^{-t}$$

$$+ \sum_{k=1}^{n} [u_{13,k,av} + v_{1,k,av} + v_{2,k,av}] \tag{1.51}$$

$$\dot{u}_{13,k,av} + p_k u_{13,k,av} = 0, \quad \dot{v}_{1,k,av} + p_k v_{1,k,av} = 0$$

$$\dot{v}_{2,k,av} + p_k v_{2,k,av} = 0, \quad \dot{v}_{3,j,av} + p_j v_{3,j,av} = 0$$

in the original time-scale. As all of the poles p_k for all k and p_j for all j are asymptotically stable (from asymptotic stability of $H_o(s)$ and $\frac{1}{1+L(s)}$), all of the terms on the right hand side of Eqn. (1.51) for $\tilde{\theta}_{av}$ are exponentially decaying, i.e., we have

$$\tilde{\theta}_{av} = \frac{1}{s + p_0} [\epsilon^{-t}], \tag{1.52}$$

which decays to zero because $\frac{1}{s+p_0}$ is asymptotically stable by asymptotic stability of $\frac{1}{1+L(s)}$. Hence, by a standard averaging theorem such as Theorem 8.3

---

[4]Note that $A(t)$ here is different from $\mathbf{A}(t)$ in the proof of Proposition 1.5.

in Khalil [64], we see that if ω, a, ϕ, $C_i(s)$ and $C_o(s)$ are such that $\frac{1}{1+L(s)}$ is asymptotically stable, and ω is sufficiently large relative to other parameters of the state-space representation, solutions starting from small initial conditions converge exponentially to a periodic solution in an $O(1/\omega)$ neighborhood of zero. Hence, the output $\tilde{\theta}(t)$ goes to a periodic solution $\tilde{\theta}_{per}(t) = O(1/\omega)$. We now proceed to put the system in the standard form for singular perturbation analysis through making the transformation $\delta\tilde{\theta} = \tilde{\theta}(t) - \tilde{\theta}_{per}(t)$ in Eqns. (1.34), (1.35) and get:

$$\text{den}\{H_i(s)\}[\delta\tilde{\theta} + \tilde{\theta}_{per}]$$
$$= \epsilon^{-t} + \text{num}\{H_i(s)\}\left[u_{13}(t) + \epsilon^{-t} - f''\sin(\omega t - \phi))y'_{osp}\right]$$
$$y'_{osp} = (1 + H^{sp}_{obp}(s))[y_{osp}] \tag{1.53}$$
$$y_{osp} = H_{osp}(s)\left[\theta_0(\delta\tilde{\theta} + \tilde{\theta}_{per})\right]. \tag{1.54}$$

By linearity of the system described by the Eqns. (1.53), (1.54), we have that the reduced model in the new coordinates (replacing $H_{osp}(s)$ with its unity static gain) is given by

$$\text{den}\{H_i(s)\}[\delta\tilde{\theta}]$$
$$= -f''\text{num}\{H_i(s)\}\left[\sin(\omega t - \phi))(1 + H^{sp}_{obp}(s))[\theta_0\delta\tilde{\theta}]\right], \tag{1.55}$$

which has the state space representation

$$\dot{\mathbf{x}} = A(t)\mathbf{x}; \quad \delta\tilde{\theta} = x_1, \tag{1.56}$$

where $A(t)$ is the same as in Eqn. (1.47). Hence $\delta\tilde{\theta}$ converges exponentially to the origin. This shows that the reduced model is exponentially stable. From exponential stability of $H_{osp}(s)$, we have exponential stability of the boundary layer model

$$\frac{d\mathbf{y}}{d\tau} = \mathbf{A}_{osp}\mathbf{y}, . \tag{1.57}$$

Hence, by the Singular Perturbation Lemma A.3, we have that in the overall unreduced system in Eqns. (1.53), (1.54), the solution converges to an $O(1/M)$ neighborhood of the origin. Hence, $\delta\tilde{\theta}(t)$ converges to a $O(1/M)$ neighborhood of the origin. Therefore, $\tilde{\theta}$ converges exponentially to a $O(1/\omega) + O(1/M) = O(\delta)$ neighborhood of the origin. Further, the output error $\tilde{y}$ decays to $O\left(a^2 + \delta^2\right)$:

$$\tilde{y} = F_o(s)\left[\frac{f''}{2}(\theta - \theta^*)^2\right] = F_o(s)\left[\frac{f''}{2}(\tilde{\theta} - \theta_0)^2\right] = O(a^2 + \delta^2), \tag{1.58}$$

which completes the proof.   Q.E.D.

From Eqn. (1.39), we notice that $C_i(s)$ appears *linearly* in $L(s)$ (through $H_i(s) = C_i(s)\Gamma_\theta(s)F_i(s)$). This property allows systematic design using *linear control tools*. The conditions of Theorem 1.8 motivate the steps of a design algorithm below.

1.2.3 Single Parameter Compensator Design

In the design guidelines that follow, we set $\phi = 0$ which can be used separately for fine-tuning.

Algorithm 1.2.1 (Single Parameter Extremum Seeking)

1. *Select the perturbation frequency ω sufficiently large and not equal to any frequency in noise, and with $\pm j\omega$ not equal to any imaginary axis zero of $F_i(s)$.*

2. *Set perturbation amplitude a so as to obtain small steady state output error $\tilde{y}$.*

3. *Design $C_o(s)$ asymptotically stable, with zeros of $\Gamma_f(s)$ that are not asymptotically stable as its zeros, and such that $\frac{C_o(s)}{\Gamma_f(s)}$ is proper. In the case where dynamics in $F_o(s)$ are slow and strictly proper, use as many fast poles in $C_o(s)$ as the relative degree of $F_o(s)$, and as many zeros as needed to have zero relative degree of the slow part $H_{obp}(s)$ to satisfy Assumption 1.6.*

4. *Design $C_i(s)$ by any linear SISO design technique such that it does not include poles of $\Gamma_\theta(s)$ that are not asymptotically stable as its zeros, $C_i(s)\Gamma_\theta(s)$ is proper, and $\frac{1}{1+L(s)}$ is asymptotically stable.*

We examine these design steps in detail:

Step 1: Since the averaging assumption is only qualitative, we may be able to choose ω only slightly larger than the plant time constants, as we show in the examples in Sections 1.3 and 2.3. Choice of ω equal to a frequency component persistent in the noise n can lead to a large steady state tracking error $\tilde{\theta}$. In fact, our analysis can be adapted to include a bounded noise signal satisfying $\lim_{T\to\infty} \frac{1}{T}\int_0^T n\sin\omega t\, dt = 0$. Finally, if $\pm j\omega$ is a zero of $F_i(s)$, the sinusoidal forcing will have no effect on the plant.

Step 2: The perturbation amplitude a should be designed such that $a|F_i(j\omega)|$ is small; a itself may have to be large so as to produce a measurable variation in the plant output.

Step 3: In general, this design step will need designing a biproper $\frac{C_o(s)}{\Gamma_f(s)}$ when we have a slow and strictly proper $F_o(s)$ in order to satisfy Assumption 1.6. The use of fast poles in $\frac{C_o(s)}{\Gamma_f(s)}$ raises a possibility of noise deteriorating the feedback; however, the demodulation coupled with the integrating action of the input compensator prevents frequencies other than that of the forcing from entering into the feedback. While we have used the gain k in analysis to satisfy Assumption 1.6, this is not strictly necessary in design.

Step 4: We see from Algorithm 1.2.1 that $C_i(s)$ has to be designed such that $C_i(s)\Gamma_\theta(s)$ is proper; hence, for example, if $\Gamma_\theta(s) = \frac{1}{s^2}$, an improper $C_i(s) = 1+d_1 s+d_2 s^2$ is permissible. In the interest of robustness, it is desirable to design $C_i(s)$ and $C_o(s)$ to ensure minimum relative degree of $C_i(s)\Gamma_\theta(s)$ and $\frac{C_o(s)}{\Gamma_f(s)}$. This will help to provide lower loop phase and thus better phase margins. Simplification of the design for $C_i(s)$ is achieved by setting $\phi = -\angle(F_i(j\omega))$, and obtaining

$$L(s) = \frac{af''|F_i(j\omega)|}{4}H_i(s).$$

The attraction of extremum seeking is its ability to deal with uncertain plants. In our design, we can accommodate uncertainties in f'', $F_o(s)$, and $F_i(s)$, which appear as uncertainties in $L(s)$. Methods for treatment of these uncertainties are dealt with in texts such as [123]. Here we only show how it is possible to ensure robustness to variations in f''. Let $\widehat{f''}$ denote an a priori estimate of f''. Then we can represent $\frac{1}{1+L(s)}$ as $\frac{1}{1+L(s)} = \frac{1}{1+\left(1+\frac{\Delta f''}{f''}\right)P(s)}$,

where $\Delta f'' = f'' - \widehat{f''}$, and $P(s) = \frac{\widehat{f''}}{f''}L(s)$, which is at our disposal because f'' in $P(s)$ gets cancelled by f'' in $L(s)$. We design $C_i(s)$ to minimize $\|\frac{P}{1+P}\|_{H_\infty}$ which maximizes the allowable $\Delta f'' < \widehat{f''}/\|\frac{P}{1+P}\|_{H_\infty}$ under which the system is still asymptotically stable.

1.3   Single Parameter Example

To illustrate the power of the method, we present a difficult toy benchmark problem. Bounded noise n is present in all simulations. Simulation results are plotted with $f^*(t), \theta^*(t)$ in dotted lines and y, θ in solid lines.

Benchmark Example.   We use $F_i(s) = \frac{s-1}{s^2+3s+2}$, $F_o(s) = \frac{1}{s+1}$, $f(\theta) = f^*(t) + (\theta - \theta^*(t))^2$, where $f^*(t) = 0.01u(t-10)$, giving $\lambda_f\Gamma_f(s) = \frac{0.01e^{-10s}}{s}$, $\theta^*(t) = 0.01e^{0.01t}$, giving $\lambda_\theta\Gamma_\theta(s) = \frac{0.01}{s-.01}$, and $u(t-\tau)$ denotes the unit step at time τ. The benchmark example provides all the elements of difficulty that a design can face: a large relative degree of $F_i(s)F_o(s)$; unstable $\Gamma_\theta(s)$; non-minimum phase $F_i(s)$.

Using Algorithm 1.2.1 for the design, we set $\omega = 5$ rad/sec, $a = .05$, $\frac{C_o(s)}{\Gamma_f(s)} = \frac{s}{s+5}$, $\phi = -\angle(F_i(j5)) = .7955$, and $C_i(s)\Gamma_\theta(s) = 107.7\frac{s-4}{s-.01}$.

The design attains the desired goal of output minimization. The response in Figure 1.3 shows a slow transient and noise sensitivity in the parameter tracking. But, we note that because of the attenuation of the high frequency perturbation in the plant, the output y tracks the minimum $f^*(t)$ well. We note that we used $C_o(s) = 1/(s+5)$– a fast pole– as the pole in $F_o(s)$ is slow, and $F_o(s)$ is strictly proper.

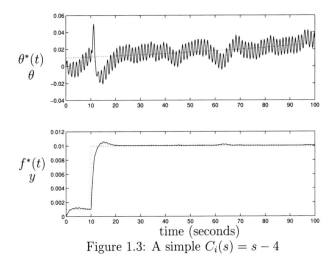

$\theta^*(t)$
θ

$f^*(t)$
y

time (seconds)

Figure 1.3: A simple $C_i(s) = s - 4$

Notes and References

Several books in the 1950-60s have been exclusively or partly dedicated to extremum seeking, including Tsien [109] (1954), Feldbaum [39] (1959), Krasovskii [66] (1963), Wilde [119] (1964), and Chinaev [29] (1969). Among the surveys on extremum seeking, we find the one by Sternby [105] particularly useful, as well as Section 13.3 in Astrom and Wittenmark [7] which puts extremum control among the most promising future areas for adaptive control. Among the many applications of extremum control overviewed in [105] and [7] are combustion processes (for IC engines, steam generating plants, and gas furnaces), grinding processes, solar cell and radio telescope antenna adjustment to maximize the received signal, and blade adjustment in water turbines and wind mills to maximize the generated power.

Extremum seeking witnessed a resurgence of interest after the publication of the stability studies in [70] and [67]. Several analyses of extremum seeking schemes [16, 37, 96, 113, 122] were presented at ACC 2000. The need for rigorous design guidelines guaranteeing performance was strongly felt in [67, 96]. Besides, numerical optimization based extremum seeking methods were used successfully. Extremum seeking via triangular search as in Zhang [122], was employed to attenuate combustor thermoacoustic oscillations and minimize diffuser losses at United Technologies Research Center (UTRC); extremum seeking via sliding modes, introduced by Korovin and Utkin [65], and analyzed and applied by Ozguner and co-workers [35, 51] and others [28, 106] on a variety of automotive problems. Even more recently, extremum seeking has been used for control of electromechanical valves [93], and for optimization of IC engine cam timing [94].

The first stability analysis of extremum seeking for a general nonlinear plant was developed in [70]. In [67], dynamic compensation was proposed for providing stability guarantees and fast tracking of changes in plant operating conditions for single parameter extremum seeking. The result in [5], on which this chapter is based, supplies the systematic design, and moreover, limits adaptation speed only by the speed of the plant dynamics.

Chapter 2

Multiparameter Extremum Seeking

Many problems that require feedback optimization are inherently multivariable: cancellation of oscillations, formation flight, and minimization of airframe drag are a few examples of problems whose practical solution will be enabled by the ability to systematically design multivariable extremum seeking schemes. This chapter supplies this need and treats the analysis based design of multivariable extremum seeking schemes:

1. It provides a multiparameter extremum seeking scheme for general time-varying extrema (Section 2.1).

2. Derives a stability test in a simple SISO format (Section 2.1).

3. Presents a systematic design algorithm based on standard LTI control techniques to satisfy the stability test (Section 2.2).

4. Supplies an analytical quantification of the level of design difficulty in terms of the number of parameters and in terms of the shape of the unknown equilibrium map (Section 2.2).

Section 2.3 presents simulation examples that compare performance of the multiparameter design and its three variations presented in Section 2.2, first for step changes, and then for more general (linear dynamic) variations in the extremum.

2.1 Output Extremization in Multiparameter Extremum Seeking

Figure 2.1 shows the multiparameter extremum seeking scheme. Analogous to the single parameter case in Section 1.2, we let $f(\theta)$ be a function of the form:

$$f(\theta) = f^*(t) + (\theta - \theta^*(t))^T \mathbf{P}(\theta - \theta^*(t)), \qquad (2.1)$$

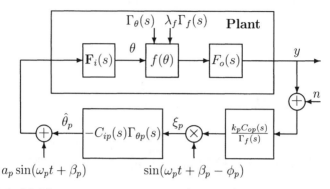

Figure 2.1: Multiparameter extremum seeking with $p = 1, 2, \ldots, l$. For p odd, $\omega_{p+1} = \omega_p$, $\beta_p = 0$, and $\beta_{p+1} = \pi/2$.

where $\mathbf{P}_{l \times l} = \mathbf{P}^T > 0$, $\theta = [\theta_1 \ldots \theta_l]^T$, $\theta^*(t) = [\theta_1^*(t) \ldots \theta_l^*(t)]^T$, $\mathcal{L}\{\theta^*(t)\} = \Gamma_\theta(s) = [\lambda_1 \Gamma_{\theta 1}(s) \ldots \lambda_l \Gamma_{\theta l}(s)]^T$, and $\mathcal{L}\{f^*(t)\} = \lambda_f \Gamma_f(s)$. Any vector function $f(\theta)$ with a quadratic minimum at θ^* can be approximated by Eqn. (2.1). In seeking maxima, i.e., $\mathbf{P} < 0$, we only need to replace $C_{ip}(s)$ with $-C_{ip}(s)$. Further, the method need not confine itself to seeking only extrema; convergence to saddle points, and any points with zero first derivative may be attained using the designs developed here simply by setting $C_{ip}(s)$ the same sign as P_{pp}. We further note here that we propose diagonal compensation in the scheme in Figure 2.1 for two reasons: while we can obtain a multiple-input single-output (MISO) sensitivity design problem, there are no systematic means of multivariable design when there is an unknown matrix gain ($\mathbf{P}$) in the plant; use of a multiple-input multiple-output (MIMO) compensator also leads to an O(l) increase in the steady-state output deviation from the extremum.

The broad principle of using m frequencies for identification/tracking of $2m$ parameters holds here. Forcing frequencies $\omega_1 < \omega_3 < \ldots < \omega_{2m-1}$ are used, where $m = \left[\frac{l+2}{2}\right]$ ($[x]$ is the greatest integer less than x). We make assumptions analogous to the single parameter case:

Assumption 2.1 $\mathbf{F}_i(s) = [F_{i1}(s) \ldots F_{il}(s)]^T$ and $F_o(s)$ are asymptotically stable and proper.

Assumption 2.2 $\Gamma_\theta(s)$ and $\Gamma_f(s)$ are strictly proper, and $F_{ip}(s)$ does not contain zeros equal to poles in $\Gamma_{\theta p}(s)$ that are not asymptotically stable.

Assumption 2.3 $C_{ip}(s)\Gamma_{\theta p}(s)$ and $\frac{C_{op}(s)}{\Gamma_f(s)}$ are proper for all $p = 1, 2, \ldots, l$.

As in the single parameter case, the signal n in Figure 2.1 denotes measurement noise. In the multiparameter case, we make an additional assumption upon the perturbation frequencies:

Assumption 2.4 $\omega_p + \omega_q \neq \omega_r$ *for any* $p, q, r = 1, 2, \ldots, l.$

The purpose of this assumption is to preclude bias terms arising from demodulation in the case of a quadratic nonlinearity. We can always satisfy this assumption since the choice of frequencies is at our disposal. We expatiate further upon this assumption at the end of this section.

Finally, we note that if the frequencies $\omega_1, \ldots, \omega_m$, are rational, they can be written as $\omega_1, \frac{p_1}{q_1}\omega_1, \frac{p_2}{q_2}\omega_1, \ldots, \frac{p_{m-1}}{q_{m-1}}\omega_1$. The time periods in the system are $\frac{2\pi}{\omega_p}$, $\frac{2\pi}{(\omega_p - \omega_q)}, \omega_p \neq \omega_q, \frac{2\pi}{(\omega_p + \omega_q)}, \frac{2\pi}{(\omega_p + \omega_q - \omega_r)}$, and $\frac{2\pi}{(\omega_p + \omega_q + \omega_r)}$, $p, q, r = 1, \ldots, m$, all of which are rational multiples of 2π. Thus, the overall time-period of the system T can be calculated as the lowest common multiple of these time-periods. We make use of this fact in Subsections 2.1.1 and 2.1.2 below.

2.1.1   Multivariable LTV Stability Test

The equations governing the pth loop of the multiparameter scheme in Figure 2.1 are as follows:

$$y \;=\; F_o(s)\left[f^* + (\theta - \theta^*)^T \mathbf{P}(\theta - \theta^*)\right] \tag{2.2}$$

$$\theta_p \;=\; F_{ip}(s)\left[a_p \sin(\omega_p t + \beta_p) - C_{ip}(s)\Gamma_{\theta p}(s)[\xi_p]\right] \tag{2.3}$$

$$\xi_p \;=\; \sin(\omega_p t + \beta_p - \phi_p)\frac{C_{op}(s)}{\Gamma_f(s)}[y + n], \tag{2.4}$$

where

$$\beta_p = \begin{cases} 0 & , \quad p \text{ odd} \\ \frac{\pi}{2} & , \quad p \text{ even} \end{cases} \tag{2.5}$$

and, for p odd, $\omega_{p+1} = \omega_p$. The definitions of tracking error $\tilde{\theta}_p$ and output error $\tilde{y}$ analogous to the single parameter case are:

$$\tilde{\theta}_p \;=\; \theta_p^* - \theta_p + \theta_{0p}; \quad \tilde{\theta} = [\tilde{\theta}_1 \ldots \tilde{\theta}_l]^T \tag{2.6}$$

$$\theta_{0p} \;=\; F_{ip}(s)[a_p \sin(\omega_p t + \beta_p)]; \quad \theta_0^T = [\theta_{01} \ldots \theta_{0l}]^T \tag{2.7}$$

$$\tilde{y} \;=\; y - F_o(s)[f^*] = F_o(s)\left[(\theta - \theta^*)^T \mathbf{P}(\theta - \theta^*)\right]. \tag{2.8}$$

We now state our first result on multiparameter output extremization:

Proposition 2.5 (Multiparameter Extremum Seeking: LTV Test)
For the system in Figure 2.1, under Assumptions 2.1–2.4, the output error $\tilde{y}$ achieves local exponential convergence to an $O(l \sum_{p=1}^{l} a_p^2)$ neighborhood of zero provided $n = 0$ and:

1. *Perturbation frequencies $\omega_1 < \omega_3 < \ldots < \omega_{2m-1}$ are rational, and $\pm j\omega_p$ is not a zero of $F_{ip}(s)$.*

2. *Zeros of $\Gamma_f(s)$ that are not asymptotically stable are also zeros of $C_{op}(s)$, for all $p = 1, \ldots, l$.*

3. *Poles of $\Gamma_{\theta p}(s)$ that are not asymptotically stable are not zeros of $C_{ip}(s)$, for any $p = 1, \ldots, l$.*

4. *$C_{op}(s)$ are asymptotically stable for all $p = 1, \ldots, l$ and the eigenvalues of the matrix $\mathbf{\Phi}(T, 0)$ lie inside the unit circle, where T is the time period of the system and $\mathbf{\Phi}(T, 0)$ is the solution at time T of the matrix differential equation*

$$\dot{\mathbf{\Phi}} = \mathbf{A}(t)\mathbf{\Phi}(t, 0), \quad \mathbf{\Phi}(0, 0) = \mathbf{I}, \tag{2.9}$$

and $\dot{\mathbf{x}} = \mathbf{A}(t)\mathbf{x}(t, 0), \mathbf{x}(0) = \mathbf{x}_0$ is a state space representation of the LTV differential equations

$$\begin{aligned}
\mathrm{den}\{H_{in}(s)\}[\tilde{\theta}_n] &= \mathrm{num}\{H_{in}(s)\}\left[\sin(\omega_n t + \beta_n - \phi_n)\right. \\
&\quad \left. \times H_{on}(s)\left[\sum_{p=1}^{l}\sum_{q=1}^{l} 2P_{pq}\tilde{\theta}_q\theta_{0p}(t)\right]\right].
\end{aligned} \tag{2.10}$$

and

$$\begin{aligned}
H_{ip}(s) &= C_{ip}(s)\Gamma_{\theta p}(s)F_{ip}(s) \\
H_{op}(s) &= k_p\frac{C_{op}(s)}{\Gamma_f(s)}F_o(s).
\end{aligned}$$

Proof: We expand $\tilde{\theta}_n$ in Eqn. (2.6), substituting for θ_n, ξ_n, and y from Eqns. (2.3), (2.4), and (2.2) respectively and get:

$$\begin{aligned}
\tilde{\theta}_n &= \theta_n^* + H_{in}(s)\left[\sin(\omega_n t + \beta_n - \phi_n)H_{on}(s)[f^* + (\theta - \theta^*)^T\mathbf{P}(\theta - \theta^*)]\right] \\
&= \theta_n^* + H_{in}(s)\left[\sin(\omega_n t + \beta_n - \phi_n)H_{on}(s)[f^* + (\tilde{\theta} - \theta_0)^T\mathbf{P}(\tilde{\theta} - \theta_0)]\right], \tag{2.11}
\end{aligned}$$

using $\theta - \theta^* = \theta_0 - \tilde{\theta}$ from Eqn. (2.6). Here, in addition to terms encountered in the single parameter case, we have to consider linear terms and higher order terms that arise due to coupling from the quadratic form:

$$(\tilde{\theta} - \theta_0)^T\mathbf{P}(\tilde{\theta} - \theta_0) = \sum_{p=1}^{l}\sum_{q=1}^{l} P_{pq}\left(\tilde{\theta}_p\tilde{\theta}_q + \theta_{0p}\theta_{0q} - \tilde{\theta}_p\theta_{0q} - \tilde{\theta}_q\theta_{0p}\right) \tag{2.12}$$

The term containing $f^*(t)$, and $\theta_{0p}\theta_{0q}$ in Eqn. (2.11) can be simplified using Lemma A.1 as in the proof of Theorem 1.8, using Assns. 2.1, 2.2, 2.3, and asymptotic stability of $C_{on}(s)$:

$$\sin(\omega_n t + \beta_n - \phi_n)H_{on}(s)[f^* + \sum_{p=1}^{l}\sum_{q=1}^{l} P_{pq}\theta_{0p}\theta_{0q}] = w_n(t) + \epsilon^{-t}, \tag{2.13}$$

where ϵ^{-t} denotes exponentially decaying terms, and

$$
\begin{aligned}
w_n(t) \;=\; & \sum_{p=1}^{l}\sum_{q=1}^{l} P_{pq}a_p a_q \left\{ C_{pq1} \sin\left[(\omega_n + \omega_p + \omega_q)t - \mu_{pq1}\right]\right. \\
& + C_{pq2} \sin\left[(\omega_n + \omega_p - \omega_q)t - \mu_{pq2}\right] \\
& + C_{pq3} \sin\left[(\omega_n - \omega_p + \omega_q)t - \mu_{pq3}\right] \\
& \left. + C_{pq4} \sin\left[\omega_n - \omega_p - \omega_q)t - \mu_{pq4}\right]\right\},
\end{aligned}
\tag{2.14}
$$

where

$$
C_{pq1} = C_{pq4} = |F_{ip}(j\omega_p)F_{iq}(j\omega_q)H_{on}(j(\omega_p + \omega_q))| \tag{2.15}
$$
$$
C_{pq2} = C_{pq3} = |F_{ip}(j\omega_p)F_{iq}(j\omega_q)H_{on}(j(\omega_p - \omega_q))|, \tag{2.16}
$$

and the constants μ_{pqr}, $r = 1,\dots,4$ depend upon ϕ_n and the phases of $F_{ip}(j\omega_p)$, $F_{iq}(j\omega_q)$, and $H_{on}(j(\omega_p \pm \omega_q))$. The function $w_n(t)$ is of order $O(\sum_{p=1}^{l}\sum_{q=1}^{l} a_p a_q) = O(l \sum_{p=1}^{l} a_p^2)$ and does not contain constant terms since from Assn. 2.4, $\omega_p + \omega_q \neq \omega_r$ for any $p, q, r = 1, \dots, l$. Using Eqns. (2.12), (2.13), and symmetry of $\mathbf{P}$, we can now rewrite Eqn. (2.11) as follows after dropping the second order terms[1]:

$$
\begin{aligned}
\tilde{\theta}_n \;=\; & \theta_n^* + H_{in}(s)\left[\sin(\omega_n t + \beta_n - \phi_n)H_{on}(s)\left[\sum_{p=1}^{l}\sum_{q=1}^{l} 2P_{pq}\tilde{\theta}_q \theta_{0p}\right]\right. \\
& \left. + w_n(t) + \epsilon^{-t}\right].
\end{aligned}
\tag{2.17}
$$

Multiplying both sides in Eqns. (2.17) with $\mathrm{den}\{H_{in}(s)\}$ gives the following set of LTV differential equations:

$$
\begin{aligned}
\mathrm{den}\{H_{in}(s)\}[\tilde{\theta}_n] = \;& \epsilon^{-t} \\
+ \mathrm{num}\{H_{in}(s)\} & \left[\sin(\omega_n t + \beta_n - \phi_n)H_{on}(s)\left[\sum_{p=1}^{l}\sum_{q=1}^{l} 2P_{pq}\tilde{\theta}_q \theta_{0p}\right]\right. \\
+ w_n(t) + \epsilon^{-t}\Big].
\end{aligned}
\tag{2.18}
$$

The θ_n^* term drops out or becomes an exponentially decaying term when operated upon by $\mathrm{den}\{\Gamma_{\theta n}(s)\}$ contained in $\mathrm{den}\{H_{in}(s)\}$. We now write a state space representation of the LTV system in Eqn. (2.10), which is the homogenous part of the LTV system in Eqn. (2.18):

$$
\dot{\mathbf{x}} = \mathbf{A}(t)\mathbf{x}, \quad \mathbf{A}(t + T) = \mathbf{A}(t). \tag{2.19}
$$

---

[1]As in the single parameter case, this is justified by Lyapunov's first method, as we have already written the system in terms of error variables $\tilde{\theta}_q$, $Q = 1, \dots, l$ thus transforming the problem to stability of the origin.

The system has a state transition matrix $\mathbf{\Phi}(t,0)$ given by the solution of

$$\dot{\mathbf{\Phi}} = \mathbf{A}(t)\mathbf{\Phi}(t,0), \ \ \mathbf{\Phi}(0,0) = \mathbf{I}. \tag{2.20}$$

The system is exponentially stable if the eigenvalues of the matrix $\mathbf{\Phi}(T,0)$ (numerically calculated above) lie within the unit circle by Property 5.11 in [97]. As the persistent part of the non-homogenous forcing term in Eqn. (2.18) is $O(l\sum_{p=1}^{l} a_p^2)$, we have convergence[2] of $\tilde{\theta}$ to $O(l\sum_{p=1}^{l} a_p^2)$ and therefore the convergence of $\tilde{y} = y - F_o(s)[f^*(t)] = F_o(s)\left[(\tilde{\theta} - \theta_0)^T \mathbf{P}(\tilde{\theta} - \theta_0)\right]$ to $O(l\sum_{p=1}^{l} a_p^2)$. Q.E.D.

2.1.2   Multivariable LTI Stability Test

As in the single parameter case, we provide a result permitting systematic design under an additional assumption (which can always be satisfied by design). To this end, we introduce the following notation:

$$H_{op}(s) = k_p \frac{C_{op}(s)}{\Gamma_f(s)} F_o(s) \overset{\triangle}{=} H_{osp,p}(s) H_{obp,p}(s)$$

$$\overset{\triangle}{=} H_{osp,p}(s)(1 + H_{obp,p}^{sp}(s)) \tag{2.21}$$

where $H_{osp,p}(s)$ denotes the strictly proper part of $H_{op}(s)$ and $H_{obp,p}(s)$ its biproper part, each of the k_p is chosen to ensure

$$\lim_{s \to 0} H_{osp,p}(s) = 1. \tag{2.22}$$

We now present the multiparameter analog of Assumption 1.6:

Assumption 2.6 *Let the smallest in absolute value among the real parts of all of the poles of $H_{osp,p}(s)$ for all p be denoted by a. Let the largest among the moduli of all of the poles of $F_{ip}(s)$ and $H_{obp,p}(s)$ for all p, be denoted by b. The ratio $M = a/b$ is sufficiently large.*

With this assumption, we separate the slow and fast dynamics in the LTV system in Eqn. (2.18) as

$$\mathrm{den}\{H_{in}(s)\}[\tilde{\theta}_n] = \epsilon^{-t}$$
$$+\mathrm{num}\{H_{in}(s)\}\left[\sin(\omega_n t + \beta_n - \phi_n)y'_{osp,n} + w_n(t) + \epsilon^{-t}\right]$$
$$y'_{osp,n} = (1 + H_{obp,n}^{sp}(s))[y_{osp,n}] \tag{2.23}$$

$$y_{osp,n} = H_{osp,n}(s)\left[\sum_{p=1}^{l}\sum_{q=1}^{l} 2P_{pq}\tilde{\theta}_q\theta_{0p}\right], \tag{2.24}$$

---

[2]Exponential stability of the homogenous part of the linear system in Eqn. 2.18 implies $\mathcal{L}$-stability of the system with forcing. Boundedness of the inhomogenous forcing yields boundedness of the solution of the forced system as in Example 6.3 in Khalil [64], and the bound on the forcing gives the order of $\tilde{\theta}$.

and write the following state-space representation for the fast dynamics $H_{osp,p}(s)$ acting on the term $\left[\sum_{p=1}^{l}\sum_{q=1}^{l}2P_{pq}\tilde{\theta}_q\theta_{0p}\right]$:

$$\frac{1}{M}\dot{\mathbf{x}}_{osp,n} = \mathbf{A}_{osp,n}\mathbf{x}_{osp,n} + \mathbf{B}_{osp,n}\left[\sum_{p=1}^{l}\sum_{q=1}^{l}2P_{pq}\tilde{\theta}_q\theta_{0p}\right] \qquad (2.25)$$

$$y_{osp,n} = \mathbf{C}_{osp,n}\mathbf{x}_{osp,n},$$

where the eigenvalues of $M\mathbf{A}_{osp,n}$ are the poles of $H_{osp,n}(s)$. Reduction of the fast dynamics in Eqn. (2.25) by singular perturbation yields the substitution

$$y_{osp,p} = \mathbf{C}_{osp,p}\mathbf{A}_{osp,p}^{-1}\mathbf{B}_{osp,p}\left[\sum_{p=1}^{l}\sum_{q=1}^{l}2P_{pq}\tilde{\theta}_q\theta_{0p}\right] = \left[\sum_{p=1}^{l}\sum_{q=1}^{l}2P_{pq}\tilde{\theta}_q\theta_{0p}\right], \qquad (2.26)$$

using Eqn. (2.22) to form a reduced order model. The purpose of this assumption is to use a singular perturbation reduction of the output dynamics and provide the LTI SISO stability test of the following theorem. As in Assumption 1.6 in Chapter 1 the assumption is made upon $H_o(s)$ for generality. As in the discussion after Assumption 1.6, its purpose is to deal with the slow strict proper part of $F_o(s)$, and perform systematic design irrespective of the speed of poles in $F_o(s)$, or its relative degree. As in the single parameter case, we make an additional assumption:

Assumption 2.7 $H_{ip}(s)$ *is strictly proper for all* p.

The theorem below also holds without this assumption, which has been made for the same reason as Assumption 1.7–brevity and clarity of the proof.

Theorem 2.8 (Multiparameter Extremum Seeking) *For the system in Figure 2.1, under Assumptions 2.1–2.7, the output error $\tilde{y}$ achieves local exponential convergence to an $O(\Delta^2 + l\sum_{p=1}^{l}a_p^2)$ neighborhood of zero, where $\Delta = 1/\omega_1 + 1/M$, provided $n = 0$ and:*

1. *Perturbation frequencies $\omega_1 < \omega_3 < \ldots < \omega_{2m-1}$ are rational, sufficiently large, and $\pm j\omega_p$ is not a zero of $F_{ip}(s)$.*

2. *Zeros of $\Gamma_f(s)$ that are not asymptotically stable are also zeros of $C_{op}(s)$, for all $p = 1, \ldots, l$.*

3. *Poles of $\Gamma_{\theta p}(s)$ that are not asymptotically stable are not zeros of $C_{ip}(s)$, for any $p = 1, \ldots, l$.*

4. *$C_{op}(s)$ are asymptotically stable for all $p = 1, \ldots, l$ and $\frac{1}{\det(\mathbf{I}_l + \mathbf{X}(s))}$ is asymptotically stable, where $X_{pq}(s)$ denote the elements of $\mathbf{X}(s)$ and*

$$X_{pq}(s) = P_{pq}a_pL_p(s) + P_{p+\delta,q}a_{p+\delta}M_p(s), \quad q = 1, \ldots, l \quad (2.27)$$

$$L_p(s) = \frac{1}{4}H_{ip}(s)\text{Re}\{e^{j\phi_p}F_{ip}(j\omega_p)\} \qquad (2.28)$$

$$M_p(s) = \frac{1}{4}H_{ip}(s)\text{Re}\{e^{j(\phi_p+\delta\frac{\pi}{2})}F_{i,p+\delta}(j\omega_p)\}, \qquad (2.29)$$

where $\delta = 1$ for p odd and $\delta = -1$ for p even.

Proof. We rewrite the linearized model[3] in Eqn. (2.17) after reduction of the fast dynamics $H_{osp,n}(s)$ to its unity static gain (from Assumption 2.6 and Eqn. (2.26)) and get

$$
\tilde{\theta}_n = \theta_n^* + H_{in}(s)\left[\sin(\omega_n t + \beta_n - \phi_n)(1 + H_{obp,n}^{sp}(s))\left[\sum_{p=1}^{l}\sum_{q=1}^{l}2P_{pq}\tilde{\theta}_q\theta_{0p}\right]\right.
$$
$$
\left. + w_n(t) + \epsilon^{-t}\right]
$$
$$
= \theta_n^* + H_{in}(s)\left[\sin(\omega_n t + \beta_n - \phi_n)\left[\sum_{p=1}^{l}\sum_{q=1}^{l}2P_{pq}\tilde{\theta}_q\theta_{0p}\right]\right.
$$
$$
\left. + \sin(\omega_n t + \beta_n - \phi_n)v_n(t) + w_n(t) + \epsilon^{-t}\right] \tag{2.30}
$$
$$
v_n(t) = H_{obp,n}^{sp}(s)\left[\sum_{p=1}^{l}\sum_{q=1}^{l}2P_{pq}|F_{ip}(j\omega_p)|\sin(\omega_p t + \angle F_{ip}(j\omega_p))\tilde{\theta}_q\right]. \tag{2.31}
$$

Using Lemmas A.1, and A.2, we obtain[4]

$$
H_{in}(s)\left[\sin(\omega_n t + \beta_n - \phi_n)[2P_{pq}\tilde{\theta}_q\theta_{0p}]\right] = H_{in}(s)\left[\mathcal{T}_{npq}[\tilde{\theta}_q] - \mathcal{K}_{npq}[\tilde{\theta}_q]\right], \tag{2.32}
$$

where

$$
\mathcal{T}_{npq}[\tilde{\theta}_q] = P_{pq}a_p\left[\text{Re}\left\{e^{j((\omega_n+\omega_p)t+\beta_n-\phi_n+\beta_p)}\tilde{\theta}_q\right\}\right] \tag{2.33}
$$
$$
\mathcal{K}_{npq}[\tilde{\theta}_q] = P_{pq}a_p\left[\text{Re}\left\{e^{j((\omega_p-\omega_n)t-\beta_n+\phi_n+\beta_p)}\tilde{\theta}_q\right\}\right]. \tag{2.34}
$$

We now rewrite Eqn. (2.30) after moving terms linear in $\tilde{\theta}_q$ to the left hand side:

$$
\tilde{\theta}_n + H_{in}(s)\left[\sum_p\sum_q(\mathcal{K}_{npq} - \mathcal{T}_{npq})[\tilde{\theta}_q] - \sin(\omega_n t + \beta_n - \phi_n)v_n(t)\right]
$$
$$
= \theta_n^* + H_{in}(s)\left[w_n(t) + \epsilon^{-t}\right]. \tag{2.35}
$$

We consider below the terms $H_{in}(s)[\mathcal{K}_{npq}]$ in Eqn. (2.32) in the following cases to explicitly show the time invariant terms:

---

[3] As in the single parameter case in Chapter 1, linearization is what yields a local convergence result.

[4] Note that Eqn. (2.32) contains additional terms of the form $H_{in}(s)[\sin(\omega_n t + \beta_n - \phi_n)H_{on}(s)[\epsilon^{-t}\tilde{\theta}_q]]$ which comes from ϵ^{-t} in $\theta_{0p}(t) = a\text{Im}\{F_{ip}(j\omega_p)e^{j\omega_p t}\} + \epsilon^{-t}$. We drop this term from subsequent analysis because it does not affect closed loop stability or asymptotic performance. It can be accounted for in three ways. One is to perform averaging over an infinite time interval in which all exponentially decaying terms disappear. The second way is to treat $\epsilon^{-t}\tilde{\theta}_q$ as a vanishing perturbation via Corollary 5.4 in Khalil [64], observing that ϵ^{-t} is integrable. The third way is to express ϵ^{-t} in state space format and let $\epsilon^{-t}\tilde{\theta}_q$ be dominated by other terms in a local Lyapunov analysis.

1. $q = n$, $p = n$: We get the term familiar from the single parameter case in Eqn. (1.41), $H_{in}(s)[\mathcal{K}_{npq}] = P_{nn}a_n L_n(s)$.

2. $q = n$, $p = n + \delta$: We get $H_{in}(s)[\mathcal{K}_{npq}] = P_{n+\delta,n}a_{n+\delta} M_n(s)$.

3. $q \neq n$, $p = n$: We get a term $H_{in}(s)[\mathcal{K}_{npq}] = P_{nq}a_n L_n(s)$.

4. $q \neq n$, $p = n + \delta$: We get a term $H_{in}(s)[\mathcal{K}_{npq}] = P_{n+\delta,q}a_{n+\delta} M_n(s)$.

5. $q \neq n, p \neq n + \delta$: $\mathcal{K}_{npq}$ is time-varying.

Using the above, we can rewrite Eqn. (2.30) in a form that shows separately the time invariant terms:

$$\tilde{\theta}_n + \sum_q X_{nq}(s)[\tilde{\theta}_q]$$

$$+ H_{in}(s) \left[\sum_{p \neq n, n+\delta} \sum_q (\mathcal{K}_{npq} - \mathcal{T}_{npq}) [\tilde{\theta}_q] \right.$$

$$\left. - \sum_{p=n,n+\delta} \sum_q \mathcal{T}_{npq}[\tilde{\theta}_q] - \sin(\omega_n t + \beta_n - \phi_n)v_n(t) \right]$$

$$= \theta_n^* + H_{in}(s) \left[w_n(t) + \epsilon^{-t} \right]. \tag{2.36}$$

The rest of the proof shows how only the time invariant terms in Eqn. (2.36) need be considered for stability of the system. We now proceed to put the Eqns. (2.35) in a form suitable for applying averaging. Dividing both sides of Eqn. (2.35) with $\det(\mathbf{I} + \mathbf{X}(s))$, we get

$$\frac{1}{\det(\mathbf{I} + \mathbf{X}(s))}[\tilde{\theta}_n]$$

$$+ Y_{in}(s) \left[\sum_p \sum_q (\mathcal{K}_{npq} - \mathcal{T}_{npq}) [\tilde{\theta}_q] - \sin(\omega_n t + \beta_n - \phi_n)v_n(t) \right]$$

$$= \frac{1}{\det(\mathbf{I} + \mathbf{X}(s))}[\theta_n^*] + Y_{in}(s) \left[w_n(t) + \epsilon^{-t} \right], \tag{2.37}$$

where $Y_{in}(s) = \frac{H_{in}(s)}{\det(\mathbf{I}+\mathbf{X}(s))} = \frac{\text{num}\{Y_{in}(s)\}}{\text{num}\{\det(\mathbf{I}+\mathbf{X}(s))\}}$ is asymptotically stable because poles of $H_{in}(s)$ that are not asymptotically stable are cancelled by zeros in $\frac{1}{\det(\mathbf{I}+\mathbf{X}(s))}$ (using condition 3 of Theorem 2.8), and $\frac{1}{\det(\mathbf{I}+\mathbf{X}(s))}$ is asymptotically stable. By noting also that zeros in $\frac{1}{\det(\mathbf{I}+\mathbf{X}(s))}$ cancel poles in $\theta_n^*(s) = \lambda_\theta \Gamma_{\theta n}(s)$ that are not asymptotically stable (using condition 3 of Theorem 2.8), and using asymptotic stability of $\frac{1}{\det(\mathbf{I}+\mathbf{X}(s))}$, we get

$$\frac{1}{\det(\mathbf{I} + \mathbf{X}(s))}[\tilde{\theta}_n] + Y_{in}(s) \left[\sum_p \sum_q (\mathcal{K}_{npq} - \mathcal{T}_{npq}) [\tilde{\theta}_q] - \sin(\omega_n t + \beta_n - \phi_n)v_n(t) \right]$$

$$= \varepsilon + Y_{in}(s)\left[w_n(t)\right] \tag{2.38}$$

$$\varepsilon = \left[\frac{1}{\det(\mathbf{I} + \mathbf{X}(s))}[\theta_n^*] + Y_{in}(s)\left[\epsilon^{-t}\right]\right], \tag{2.39}$$

where ε is exponentially decaying. Now, $Y_{in}(s)$ is strictly proper, and can therefore be written as $Y_{in}(s) = \frac{1}{s+p_{n0}}Y'_{in}(s)$, where $Y'_{in}(s)$ is proper. In terms of their partial fraction expansions, we can write $Y'_{in}(s) = A_{n0} + \sum_{k=1}^{n_n}\frac{A_{nk}}{s+p_{nk}}$, and $H^{sp}_{obp,n}(s) = \sum_{j=1}^{m_n}\frac{B_{nj}}{s+p_{nj}}$. Multiplying both sides of Eqn (2.38) with $s + p_{n0}$, expanding operators $\mathcal{K}_{npq}$ and $\mathcal{T}_{npq}$, and using the partial fraction expansions, we rewrite Eqn. (2.38) as

$$\dot{\tilde{\theta}}'_n + p_{n0}\tilde{\theta}'_n$$
$$+A_{n0}\left(u_n(t) - w_n(t) - \sin(\omega_n t + \beta_n - \phi_n)v_n(t)\right)$$
$$+ \sum_{k=1}^{n_n}\left(u_{nk}(t) - w_{nk}(t) - v_{nk}(t)\right)$$
$$= (s + p_{n0})[\varepsilon] \tag{2.40}$$
$$\tilde{\theta}'_n = \frac{1}{\det(\mathbf{I} + \mathbf{X}(s))}[\tilde{\theta}_n]$$
$$u_n(t) = \sum_p \sum_q P_{pq}a_p\left(\mathrm{Re}\left\{e^{j((\omega_p-\omega_n)t - \beta_n + \phi_n + \beta_p)}\tilde{\theta}_q\right\}\right.$$
$$\left. - \mathrm{Re}\left\{e^{j((\omega_n+\omega_p)t + \beta_n - \phi_n + \beta_p)}\tilde{\theta}_q\right\}\right) \tag{2.41}$$
$$u_{nk}(t) = \frac{A_{nk}}{s + p_{nk}}[u_n(t)], \;\; w_{nk}(t) = \frac{A_{nk}}{s + p_{nk}}[w_n(t)]$$
$$v_{nk}(t) = \frac{A_{nk}}{s + p_{nk}}[\sin(\omega_n t + \beta_n - \phi_n)v_n(t)]$$
$$v_n(t) = \sum_{j=1}^{m_n}v'_{nj}(t)$$
$$v'_{nj}(t) = \frac{B_{nj}}{s + p_{nj}}\left[\sum_{p=1}^{l}\sum_{q=1}^{l}2P_{pq}|F_{ip}(j\omega_p)|\sin(\omega_p t + \angle F_{ip}(j\omega_p))\tilde{\theta}_q\right].$$

The system of Eqns. (2.40) can be written as a set of time-varying linear differential equations that can be put into the state space form[5]:

$$\dot{\mathbf{x}} = \mathbf{A}(t)\mathbf{x} + \mathbf{A}_{12}\mathbf{x}_e + \mathbf{B}\mathbf{w}(t); \quad \tilde{\theta} = \mathbf{C}\mathbf{x} + \mathbf{C}_{12}\mathbf{x}_e \tag{2.42}$$
$$\dot{\mathbf{x}}_e = \mathbf{A}_e\mathbf{x}_e, \tag{2.43}$$

where $\mathbf{w}(t)^T = [w_1(t), \ldots, w_l(t)]$, and Eqn. (2.43) is a representation for the exponentially decaying term ε. As all forcing frequencies $\omega_1, \ldots, \omega_m$, and consequently their linear combinations, are rational, there exists a period T, which

---

[5]$\mathbf{A}(t)$, $\mathbf{B}$ and $\mathbf{C}$ in this state space representation are different from those in Proposition 2.5.

is the lowest common multiple of all the time-periods in the system, such that the system in Eqn. (2.42) is T-periodic. We get Eqns. (2.42), (2.43) into the standard form for averaging by using the transformation $\tau = \omega_1 t$, and then averaging the right hand side of the equations w.r.t time from 0 to T, i.e., $\frac{1}{T}\int_0^T(\cdot)d\tau$ treating states $\mathbf{x}$, $\mathbf{x}_e$ as constant. The averaged equations are:

$$\frac{d\mathbf{x}_{av}}{d\tau} = \frac{1}{\omega_1}\left(\mathbf{A}_{av}\mathbf{x}_{av} + \mathbf{A}_{12}\mathbf{x}_{eav}\right), \ \tilde{\theta}_{av} = \mathbf{C}\mathbf{x}_{av} + \mathbf{C}_{12}\mathbf{x}_{eav} \quad (2.44)$$

$$\frac{d\mathbf{x}_{eav}}{d\tau} = \frac{1}{\omega_1}\mathbf{A}_e\mathbf{x}_{eav}, \quad (2.45)$$

where $\mathbf{A}_{av} = \frac{1}{T}\int_0^T\mathbf{A}(\tau)d\tau$. This yields

$$\begin{aligned}
&\dot{\tilde{\theta}}'_{n,av} + p_{n0}\tilde{\theta}'_{n,av} \\
&-A_{n0}\left(u_{n,av} - v_{n,av}(t)\right) + \sum_{k=1}^{n_n}\left(u_{nk,av} - w_{nk,av} - v_{nk,av}\right) \\
&= (s + p_{n0})[\varepsilon] \quad (2.46) \\
&\tilde{\theta}'_{n,av} = \frac{1}{\det(\mathbf{I} + \mathbf{X}(s))}[\tilde{\theta}_{n,av}] \\
&u_{n,av} = A_{nk}\sum_{p=n,n+\delta}\sum_q P_{pq}a_p\left(\mathrm{Re}\left\{e^{j(\phi_n - \beta_n + \beta_p)}\tilde{\theta}_{q,av}\right\}\right) \\
&\dot{u}_{nk,av}(t) + p_{nk}u_{nk,av}(t) = u_{n,av}, \ \dot{w}_{nk,av} + p_{nk}w_{nk,av} = 0 \\
&\dot{v}_{nk,av} + p_{nk}v_{nk,av} = v_{n,av} \\
&v_{n,av} = \sum_{j=1}^{m_n}v'_{nj,av} \\
&\dot{v}'_{nj,av} + p_{nj}v'_{nj,av} = 0,
\end{aligned}$$

in the original time-scale. As all of the poles p_{nk} for all n, k and p_{nj} for all n, j are asymptotically stable (from asymptotic stability of $H_{on}(s)$ for all n and $\frac{1}{\det(\mathbf{I} + \mathbf{X}(s))}$), all of the terms on the right hand side of Eqn. (2.46) for $\tilde{\theta}'_{n,av}$ are exponentially decaying for all n, i.e., we have

$$\begin{aligned}
&(s + p_{n0})\frac{1}{\det(\mathbf{I} + \mathbf{X}(s))}[\tilde{\theta}_{n,av}] \\
&+ Y'_{in}(s)\left[\sum_{p=n,n+\delta}\sum_q P_{pq}a_p\mathrm{Re}\left\{e^{j(\phi_n - \beta_n + \beta_p)}[\tilde{\theta}_{q,av}]\right\}\right] \\
&= (s + p_{n0})\frac{1}{\det(\mathbf{I} + \mathbf{X}(s))}\left[\tilde{\theta}_{n,av} + \sum_q X_{nq}(s)[\tilde{\theta}_{q,av}]\right] \\
&= (s + p_{n0})\frac{1}{\det(\mathbf{I} + \mathbf{X}(s))}[\varepsilon], \quad (2.47)
\end{aligned}$$

incorporating all exponentially decaying terms into ε. Dividing both sides of Eqn. (2.47) with $(s + p_{n0})$, we get

$$
\frac{1}{\det(\mathbf{I} + \mathbf{X}(s))} \left[\tilde{\theta}_{n,av} + \sum_q X_{nq}(s)[\tilde{\theta}_{q,av}] \right]
$$
$$
= \frac{1}{\det(\mathbf{I} + \mathbf{X}(s))}[\varepsilon], \tag{2.48}
$$

which can be rearranged to give

$$
\frac{1}{\det(\mathbf{I} + \mathbf{X}(s))} \left[\tilde{\theta}_{n,av} + \sum_q X_{nq}(s)[\tilde{\theta}_{q,av}] \right] = \frac{1}{\det(\mathbf{I} + \mathbf{X}(s))}[\theta_n^*] + Y_{in}(s) \left[\epsilon^{-t} \right]
$$
$$
= \frac{1}{\det(\mathbf{I} + \mathbf{X}(s))} \left[\theta_n^* + H_{in}(s) \left[\epsilon^{-t} \right] \right], \tag{2.49}
$$

using $Y_{in}(s) = \frac{H_{in}(s)}{\det(\mathbf{I}+\mathbf{X}(s))}$, and Eqn. 2.39 for ε. Eqn. (2.49) represents a system of equations that can be written in matrix form as

$$
\tilde{\theta}_{av} = (\mathbf{I} + \mathbf{X}(s))^{-1} \left[\theta^* + \mathbf{H}_i(s) \left[\epsilon^{-t} \right] \right], \tag{2.50}
$$

where $\mathbf{H}_i(s) = [H_{i1}(s), \ldots, H_{il}(s)]$. $\tilde{\theta}_{av}$ decays to zero because $(\mathbf{I} + \mathbf{X}(s))^{-1}$ is asymptotically stable (owing to asymptotic stability of $\frac{1}{\det(\mathbf{I}+\mathbf{X}(s))}$), and zeros in $(\mathbf{I} + \mathbf{X}(s))^{-1}$ cancel unstable poles in θ^* and $\mathbf{H}_i(s)$. Hence, by a standard averaging theorem such as Theorem 8.3 in Khalil [64], we see that if ω_p, a_p, ϕ_p, $C_{ip}(s)$ and $C_{op}(s)$ for all $p = 1, \ldots, l$ are such that $\frac{1}{\det(\mathbf{I}+\mathbf{X}(s))}$ is asymptotically stable, $C_{op}(s)$ is asymptotically stable, ω_1 is sufficiently large relative to parameters of the state-space representation, solutions starting from small initial conditions converge exponentially to a periodic solution in an $O(1/\omega_1)$ neighborhood of zero. Hence, $\tilde{\theta}$ goes to a periodic solution $\tilde{\theta}_{per} = O(1/\omega_1)$. We now proceed to put the system in the standard form for singular perturbation analysis through making the transformation $\delta\tilde{\theta} = \tilde{\theta}(t) - \tilde{\theta}_{per}(t)$ in Eqns. (2.23), (2.24) and get:

$$
\text{den}\{H_{in}(s)\}[\delta\tilde{\theta}_n + \tilde{\theta}_{per,n}] = \epsilon^{-t}
$$
$$
+ \text{num}\{H_{in}(s)\} \left[\sin(\omega_n t + \beta_n - \phi_n) y'_{osp,n} + w_n(t) + \epsilon^{-t} \right] \tag{2.51}
$$
$$
y'_{osp,n} = (1 + H^{sp}_{obp,n}(s))[y_{osp,n}]
$$
$$
y_{osp,n} = H_{osp,n}(s) \left[\sum_{p=1}^l \sum_{q=1}^l 2P_{pq}(\delta\tilde{\theta}_q + \tilde{\theta}_{per,q})\theta_{0p} \right]. \tag{2.52}
$$

By linearity of the system described by the Eqns. (2.51), (2.52), we have that the reduced model in the new coordinates (replacing $H_{osp,n}(s)$ with its unity static gain) is given by

$$\text{den}\{H_{in}(s)\}[\delta\tilde{\theta}_n + \tilde{\theta}_{per,n}] = \text{num}\{H_{in}(s)\}\left[\sin(\omega_n t + \beta_n - \phi_n)y'_{osp,n}\right]$$

$$y'_{osp,n} = (1 + H^{sp}_{obp,n}(s))\left[\sum_{p=1}^{l}\sum_{q=1}^{l} 2P_{pq}(\delta\tilde{\theta}_q + \tilde{\theta}_{per,q})\theta_{0p}\right] \qquad (2.53)$$

which has the state space representation

$$\dot{\mathbf{x}} = \mathbf{A}(t)\mathbf{x}; \quad \delta\tilde{\theta} = \mathbf{C}\mathbf{x}, \qquad (2.54)$$

where $\mathbf{A}(t)$ and $\mathbf{C}$ are the same as in Eqn. (2.42). Hence $\delta\tilde{\theta}$ converges exponentially to the origin. This shows that the reduced model is exponentially stable. From exponential stability of $H_{osp,n}(s)$, we have exponential stability of the boundary layer models

$$\frac{d\mathbf{y}_{osp,n}}{d\tau} = \mathbf{A}_{osp,n}\mathbf{y}_{osp,n}. \qquad (2.55)$$

Hence, by the Singular Perturbation Lemma A.3, we have that in the overall unreduced system in Eqns. (2.51), (2.52), the solution converges to an $O(1/M)$ neighborhood of the origin. Therefore, $\tilde{\theta}(t) - \tilde{\theta}_{per}(t)$ converges to a $O(1/M)$ neighborhood of the origin. Hence, $\tilde{\theta}$ converges exponentially to a $O(1/\omega_1) + O(1/M) = O(\Delta)$ neighborhood of the origin. Further, the output error $\tilde{y}$ decays to $O(\Delta^2 + l\sum_{p=1}^{l} a_p^2)$:

$$\begin{aligned}
\tilde{y} &= F_o(s)[(\theta - \theta^*)^T \mathbf{P}(\theta - \theta^*)] = F_o(s)[(\tilde{\theta} - \theta_0)^T \mathbf{P}(\tilde{\theta} - \theta_0)] \\
&= F_o(s)[\tilde{\theta}^T \mathbf{P}\tilde{\theta} + \theta_0^T \mathbf{P}\theta_0 - 2\tilde{\theta}^T \mathbf{P}\theta_0] \\
&= O(\Delta^2 + \sum_{p=1}^{l}\sum_{q=1}^{l} a_p a_q) = O(\Delta^2 + l\sum_{p=1}^{l} a_p^2). \qquad (2.56)
\end{aligned}$$

Q. E. D.

We have proved Theorem 2.8 for the case where a single frequency is used for tracking of two parameters. Because of the coupling this introduces through the $M_p(s)$ terms in each of the $X_{pq}(s)$, the process of multiparameter design may become difficult. When it is possible to use more frequencies, the design may be simpler. Hence, we provide here a corollary to Theorem 2.8 when a forcing frequency is dedicated to tracking only one parameter instead of two:

Corollary 2.9 *If forcing frequencies $\omega_1 < \omega_2 < \ldots < \omega_s$, $2m - 1 < s \le l$ are chosen for the scheme in Figure 2.1, and all the other conditions of Theorem 2.8 hold, its result also holds with $X_{pq}(s) = P_{pq}a_p L_p(s)$ for each p where $\omega_p \ne \omega_r$ for any $r \ne p$, and $X_{pq}(s)$ is given by Eqn (2.27) otherwise.*

The result follows from the fact that the coupling terms $M_p(s)$ vanish when a forcing frequency is used only for one tracking loop.

Now, we briefly discuss Assumption 2.4. From Eqns. (2.13), (2.14) it is clear that the assumption precludes constant terms in $w_n(t)$ only when the nonlinearity is quadratic. For a general nonlinearity, the frequencies can be designed incommensurate, and the analysis result arrived at by infinite time averaging. Even without this assumption, exponential convergence of $\tilde{y}$ to a neighborhood of the origin can be attained if the constant terms in $w_n(t)$ are small, but the analysis is longer, since we would have to linearize equations for $\tilde{\theta}_n$ about those constant terms, and then perform averaging.

2.2   Multiparameter Design

The process of design for the multiparameter case can be divided into the following sequential steps: selection of frequencies $\omega_1, \omega_2, \ldots, \omega_l$, selection of perturbation amplitudes $a_1, a_2, \ldots, a_l$, design of compensators $C_{op}(s)$ and $C_{ip}(s)$ for each p, to satisfy the conditions of Theorem 2.8.

The complexity of multiparameter design arises from the need for asymptotic stabilization of $\frac{1}{\det(\mathbf{I}+\mathbf{X}(s))}$, which is intricately coupled. Methods of decentralized control, such as those in [104], do not apply to our problem because the coupling between different subsystems enters through the compensators $C_{ip}(s)$ due to a single output being used for measurement. We propose here a method of reducing the general problem to allow independent SISO design of each of the compensators $C_{ip}(s)$. The method involves domination of the off diagonal terms in $\mathbf{I}+\mathbf{X}(s)$ by the diagonal terms, and may be termed *diagonal domination design*.

Before we proceed, we remind the reader of the notion of a *permanent* of a matrix. The permanent of a matrix $\mathbf{A}$ is defined as $\text{per}\,\mathbf{A} = \sum_\sigma \prod_{i=1}^n a_{i,\sigma(i)}$, where the sum runs over all $n!$ permutations σ of $\{1, \ldots, n\}$, and $\sigma(i)$ is the ith element of the permutation σ. We note that the permanent of a matrix is simply the sum of all the terms in its determinant, with all the products $\prod_{i=1}^n a_{i,\sigma(i)}$ entering with coefficient 1 instead of a power of -1.

Proposition 2.10 *Let ρ_l^* denote the unique solution in the interval $(0, 1]$ of the polynomial equation* $\text{per}\,(\Sigma(\rho)) = 2$, $\Sigma(\rho) = \begin{pmatrix} 1 & \rho & \cdots & \rho \\ \rho & 1 & \ddots & \vdots \\ \vdots & \ddots & \ddots & \rho \\ \rho & \cdots & \rho & 1 \end{pmatrix}_{l \times l}$. *If*

$\frac{X_{pp}(s)}{1+X_{pp}(s)}$ *are asymptotically stable and* $\|\frac{X_{pq}}{1+X_{pp}}\|_{H_\infty} < \rho_l^*$ *for all $p \neq q$, where $X_{pq}(s)$ are defined in Eqn. (2.27), then* $\frac{1}{\det(\mathbf{I}+\mathbf{X}(s))}$ *is asymptotically stable.*

From the definition of the permanent of a matrix, $\text{per}\,(\Sigma(\rho))$ is a polynomial

Table 2.1: Design difficulty in general design increases with dimension l

l	2	3	4	5	6	7	8	9	10
ρ_l^*	1	0.5	0.3239	0.2367	0.1855	0.1520	0.1284	0.1111	0.0978

with positive integer coefficients and thus a monotonically increasing function of ρ when $\rho > 0$. Since $\mathrm{per}\,(\Sigma(0)) = 1$ and $\mathrm{per}\,(\Sigma(1)) \geq 2$, we have by continuity, a unique solution to the equation $\mathrm{per}\,(\Sigma(\rho)) = 2$ in the interval $(0, 1]$. The equation $\mathrm{per}\,(\Sigma(\rho)) = 2$ expands as $\rho^2 = 1$ and $2\rho^3 + 3\rho^2 = 1$ in two and three dimensions, respectively, yielding $\rho_2^* = 1$, and $\rho_3^* = 0.5$. Thus the crux of Proposition 2.10 is that if the transfer functions $\frac{X_{pq}(s)}{1+X_{pp}(s)}$ are norm bounded by a number ρ_l^* that *depends only upon the dimension of the problem* l, we have asymptotic stability of $\frac{1}{\det(\mathbf{I}+\mathbf{X}(s))}$. For convenience, we list values of ρ_l^* upto $l = 10$ in Table 2.1. It can be shown that $\frac{1}{\rho_l^*} \leq \sqrt{l! - 1}$.

Proof of Proposition 2.10: We first rewrite the determinant of $(\mathbf{I} + \mathbf{X}(s))$ as follows:

$$\det(\mathbf{I} + \mathbf{X}(s)) = \det \begin{pmatrix} 1 & \frac{X_{12}(s)}{1+X_{11}(s)} & \frac{X_{13}(s)}{1+X_{11}(s)} & \cdots & \frac{X_{1l}(s)}{1+X_{11}(s)} \\ \frac{X_{21}(s)}{1+X_{22}(s)} & 1 & \frac{X_{23}(s)}{1+X_{22}(s)} & \cdots & \frac{X_{2l}(s)}{1+X_{22}(s)} \\ \vdots & & \ddots & \ddots & \vdots \\ \frac{X_{l1}(s)}{1+X_{ll}(s)} & \frac{X_{l2}(s)}{1+X_{ll}(s)} & \frac{X_{l3}(s)}{1+X_{ll}(s)} & \cdots & 1 \end{pmatrix}$$

$$\times \prod_{p=1}^{l}(1 + X_{pp}(s)) \tag{2.57}$$

$$= \det(\mathbf{Y}(s)) \prod_{p=1}^{l}(1 + X_{pp}(s)) \tag{2.58}$$

$$= (1 + W(s)) \prod_{p=1}^{l}(1 + X_{pp}(s)), \tag{2.59}$$

where Eqns. (2.57), (2.58), and (2.59) define $\mathbf{Y}(s)$ and $W(s)$. Therefore, we have

$$\frac{1}{\det(\mathbf{I} + \mathbf{X}(s))} = \frac{1}{(1 + W(s))\prod_{p=1}^{l}(1 + X_{pp}(s))}. \tag{2.60}$$

Now, we note that as $\frac{X_{pp}(s)}{1+X_{pp}(s)} \in H_\infty$ for each p, each of $\frac{1}{1+X_{pp}(s)}$ is asymptotically stable. Hence, if we can achieve $\|W\|_{H_\infty} < 1$, we have asymptotic stability of $\frac{1}{\det(\mathbf{I}+\mathbf{X}(s))}$. Using $1 + W(s) = \det(\mathbf{Y}(s))$, we have:

$$W(s) = \sum_{\sigma} \mathrm{sgn}\sigma \prod_{i=1}^{l} y_{i,\sigma(i)}(s), \tag{2.61}$$

where the sum runs over $l! - 1$ permutations σ of $\{1, \ldots, l\}$ excluding the permutation $\{1, \ldots, l\}$ to account for the cancellation of unity in Eqn. (2.61), sgnσ is positive or negative depending upon whether the number of pairwise interchanges needed to arrive at the permutation σ from the permutation $\{1, \ldots, l\}$ is even or odd, and $\sigma(i)$ is the ith element of the permutation σ.

We are now in a position to bound the H_∞ norm of $W(s)$ through repeated application of the triangle inequality and the submultiplicative property of the H_∞ norm:

$$\|W\|_{H_\infty} \leq \sum_\sigma \| \prod_{i=1}^{l} y_{i,\sigma(i)}(s)\|_{H_\infty} \leq \sum_\sigma \prod_{i=1}^{l} \|y_{i,\sigma(i)}(s)\|_{H_\infty}. \qquad (2.62)$$

Substituting $\|\frac{X_{pq}}{1+X_{pp}}\|_{H_\infty} < \rho_l^*$ for all p, q, in Eqn. (2.62), and using the fact that ρ_l^* is the unique solution of the equation per$(\Sigma(\rho)) = 2$ in the interval $(0, 1]$, we have

$$\|W\|_{H_\infty} < \text{per}\Sigma(\rho_l^*) - 1 = 1. \qquad (2.63)$$

From asymptotic stability of each of $\frac{1}{1+X_{pp}(s)}$, and of $\frac{1}{1+W(s)}$, we have asymptotic stability of $\frac{1}{\det(\mathbf{I}+\mathbf{X}(s))}$ from Eqn. (2.60).   Q.E.D.

While Proposition 2.10 provides a sufficient condition for asymptotic stability of $\frac{1}{\det(\mathbf{I}+\mathbf{X}(s))}$, it does not provide means to guarantee it. Hence the problem has now to be transformed to permit systematic design of the compensators $C_{ip}(s)$ to achieve $\|\frac{X_{pq}}{1+X_{pp}}\|_{H_\infty} < \rho_l^*$ for all p, q. To this end, we express the off diagonal terms of $\mathbf{X}(s)$ as perturbations of the diagonal terms in the case where different forcing frequencies $\omega_1 < \omega_2 < \ldots < \omega_l$ are chosen for each of the parameter tracking loops. In this case, from Corollary 2.9, we have

$$X_{pq}(s) = P_{pq}a_p L_p(s) \qquad (2.64)$$

because the coupling terms $M_p(s)$ do not arise. Thus, we have

$$\frac{X_{pq}(s)}{1 + X_{pp}(s)} = \frac{P_{pq}}{P_{pp}} \frac{X_{pp}(s)}{1 + X_{pp}(s)}. \qquad (2.65)$$

Taking the H_∞ norm of both sides of Eqn. (2.65), and using the submultiplicative property of the H_∞ norm, we get the following corollary to Proposition 2.10:

Theorem 2.11 *Consider the system from Theorem 2.8 with separate forcing frequencies $\omega_1 < \omega_2 < \ldots < \omega_l$ for each of the parameter tracking loops. If $\frac{X_{pp}(s)}{1+X_{pp}(s)}$ are asymptotically stable and $|P_{pq}| < \frac{\rho_l^*}{\|\frac{X_{pp}}{1+X_{pp}}\|_{H_\infty}} P_{pp}$ for each $q \neq p$, then $\frac{1}{\det(\mathbf{I}+\mathbf{X}(s))}$ is asymptotically stable.*

Hence, we can design $C_{ip}(s)$ to minimize $\|\frac{X_{pp}}{1+X_{pp}}\|_{H_\infty}$ for each p, which maximizes the allowable $|\frac{P_{pq}}{P_{pp}}|$. Diagonal dominance in a positive definite matrix $\mathbf{P}$ simply means that the coordinate axes of the level surfaces $(\theta - \theta^*)^T \mathbf{P}(\theta - \theta^*)$ are close to the principal axes in orientation. The need for dominance of diagonal terms in the Hessian of the nonlinearity $\mathbf{P}$ thus has a simple geometric interpretation: the inputs θ should be close to being along the principal axes of the level surfaces of the nonlinearity. Clearly, the difficulty of control design increases with dimension as ρ_l^* decreases roughly as $1/l$. For high dimensions, the problem may not have a solution. For the important case of optimizing a static map, where $F_{ip}(s) = F_o(s) = 1$, $\Gamma_{\theta p}(s) = 1/s$ for each p, and $\Gamma_f(s) = 1/s$, we can choose separate forcing frequencies for each of the parameter tracking loops, $C_{op}(s) = 1/(s+h), h > 0$, for all p, $C_{ip}(s) = k_p > 0$, and obtain $X_{pp}(s) = \frac{k_p a_p P_{pp}}{s}$. We have stability of $(\mathbf{I}+\mathbf{X}(s))^{-1}$ via a passivity argument: $\mathbf{X}(s)$ can be written as the product of a diagonal matrix of integrators and the positive definite matrix $\mathbf{P}$ of the nonlinearity, which is SPR.

To sum up the process of design, we state a multiparameter design algorithm:

Algorithm 2.2.1 (Multiparameter Design Algorithm)

1. *Select $\omega_1, \omega_2, \ldots, \omega_l$ sufficiently large, not equal to frequencies in noise, and with $\pm j\omega_p$ not equal to imaginary axis zeros of $F_{ip}(s)$.*

2. *Set perturbation amplitudes a_p so as to obtain small steady state output error $\tilde{y}$.*

3. *Design each $C_{op}(s)$ asymptotically stable, with zeros that include the zeros of $\Gamma_f(s)$ that are not asymptotically stable, and such that $\frac{C_{op}(s)}{\Gamma_f(s)}$ is proper. In the case where dynamics in $F_o(s)$ are slow and strictly proper, use as many fast fast poles in $C_{op}(s)$ as the relative degree of $F_o(s)$, and as many zeros as needed to have zero relative degree of the slow part $H_{obp,p}(s)$ to satisfy Assumption 2.6.*

4. *For each $p = 1, \ldots, l$, design $C_{ip}(s)$ such that it does not include poles of $\Gamma_{\theta p}(s)$ that are not asymptotically stable as its zeros, $C_{ip}(s)\Gamma_{\theta p}(s)$ is proper, and $\frac{1}{\det(\mathbf{I}+\mathbf{X}(s))}$ is asymptotically stable. Asymptotic stability of $\frac{1}{\det(\mathbf{I}+\mathbf{X}(s))}$ may be achieved by designing $C_{ip}(s)$ to minimize $\|\frac{X_{pp}}{1+X_{pp}}\|_{H_\infty}$ for each p, using the result in Theorem 2.11.*

We note that the theory permits the forcing frequencies to be very close, and we use close frequencies in our simulation studies in Section 2.3. Further, the condition $\omega_p + \omega_q \neq \omega_r$ for each $p, q, r = 1, \ldots, l$ used in [96] is not necessary for the output to converge to a neighborhood of the extremum, but helpful in simplifying the analysis; it ensures that the averaged Eqn. (2.44) has its equilibrium at the origin, in the case of a quadratic nonlinearity.

We can either use a separate frequency for each parameter tracking loop or use one frequency for every two parameter tracking loops, or something in between. In general, using a single frequency to force two parameter tracking loops leads to greater coupling, and consequent difficulty of design.

Design Variations. The design procedure for multiparameter extremum seeking offers theoretical guarantees of local stability and performance. The results rest upon an averaging analysis that averages out oscillatory terms in $Y_{in}(s)[w_n(t)] = O(\sum_{p=1}^{l} \sum_{q=1}^{l} a_p a_q)$ and in $Y_{in}(s) \left[\sum_p \sum_q \mathcal{T}_{npq}[\tilde{\theta}_q] - \mathcal{K}_{npq}[\tilde{\theta}_q] \right]$ (see Eqn. (2.38)). The magnitude of these oscillations can be large, and can mean a highly oscillatory output about the extremum, or even loss of stability. Here we propose design variations within the framework of the analysis above that enhance the practical utility of the design algorithm by attenuation of the oscillatory terms by a factor ϵ as they pass through the plant ($\mathbf{F}_i(s)$ and $F_o(s)$) and filters ($\frac{C_{on}(s)}{\Gamma_f(s)}$ and $C_{in}(s)\Gamma_{\theta n}(s)$):

1. Attenuation through plant dynamics, $\mathbf{F}_i(s)$ and $F_o(s)$ (High frequency design):

   (a) Select ω_1 such that $F_{in}(j\Omega) < \epsilon$ for each $n = 1, \ldots, l$, and $|F_o(j\Omega)| < \epsilon$ for all $\Omega > \omega_1$.

   (b) Choose each of the other frequencies ω_n large enough to attain $|\omega_n - \omega_p - \omega_q| \geq \omega_1$ for all $n, p, q = 1, \ldots, l$. This will yield $|F_{in}(j(\omega_n - \omega_p - \omega_q))|, |F_o(j(\omega_n - \omega_p - \omega_q))| \leq \epsilon$.

   (c) Perform steps 2, 3, 4 in Algorithm 2.2.1.

   If both $\mathbf{F}_i(s)$ and $F_o(s)$ are relative degree zero, we will not be able to achieve arbitrary attenuation ϵ.

2. Attenuation through tracking compensator $C_{in}(s)$:

   (a) Perform steps 1, 2, 3 of Algorithm 2.2.1 and write $C_{in}(s) = C'_{in}(s)F_{LPn}(s)$.

   (b) Design an asymptotically stable, minimum phase low-pass filter $F_{LPn}(s)$ such that $|F_{LPn}(j\Omega)| \leq \epsilon$ for all $\Omega > |\omega_n - \omega_p - \omega_q|$ for all $n, p, q = 1, \ldots, l$.

   (c) Design each $C'_{in}(s)$ as the $C_{in}(s)$ in step 4 of Algorithm 2.2.1 with $\Gamma_{\theta n}(s)$ replaced by $F_{LPn}(s)\Gamma_{\theta n}(s)$ with the additional constraint that poles and zeros in it do not cancel any poles or zeros of $F_{LPn}(s)$.

3. Attenuation through output compensator $C_{on}(s)$:

   (a) Perform steps 1, and 2 of Algorithm 2.2.1 and write $C_{on}(s) = C'_{on}(s)F_{BPn}(s)$.

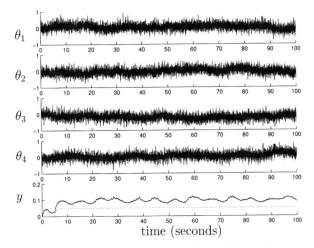

Figure 2.2: Step changes in plant: Basic design

(b) Design an asymptotically stable, minimum phase band-pass filter $F_{BPn}(s)$ such that $|F_{BPn}(j\Omega)| \le \epsilon$ for all $\Omega \ne \omega_n$, where $\Omega \in \{\omega_p, \omega_p \pm \omega_q, \omega_n \pm \omega_p \pm \omega_q\}$, $n, p, q = 1, \ldots, l$.

(c) Design each $C'_{on}(s)$ as the $C_{on}(s)$ in step 3 of Algorithm 2.2.1 with $\Gamma_f(s)$ replaced by $\frac{\Gamma_f(s)}{F_{BPn}(s)}$, with the additional constraint that poles and zeros in it do not cancel any poles or zeros of $F_{BPn}(s)$.

(d) Perform step 4 of Algorithm 2.2.1 as before.

It is clear that each one of these three design variations, whose objective is to attenuate the effect of the probing signals, will make stability more difficult to achieve.

2.3   Multiparameter Simulation Study

In this section, we simulate multiparameter extremum seeking designs generated using the methods proposed in Section 2.2. We first consider step changes in plant parameters and then more general changes. The examples we provide can be used as benchmarks for further design improvements. Simulation results are plotted with $\theta_p^*(t)$, $f^*(t)$ in dotted lines and θ_p, y in solid lines ($p = 1, \ldots, 4$). In all the examples considered below, $\phi_p = 0, p = 1, \ldots, 4$ is used.

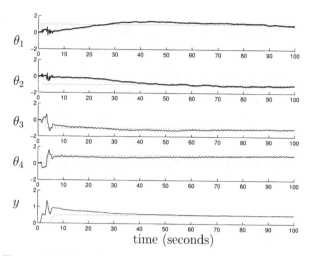

Figure 2.3: Step changes in plant: High frequency design

2.3.1   Step Variations in $\theta^*(t)$ and $f^*(t)$

The plant we consider has $F_{ip}(s) = 1$ for $p = 1,\ldots,4$, $F_o(s) = \frac{1}{s+2}$, $f(\theta) = f^*(t) + (\theta - \theta^*(t))^T \mathbf{P}(\theta - \theta^*(t))$, $P_{pq} = 1$ if $p = q$, and $P_{pq} = 0.5$ otherwise. The plant variations are $\theta_1^*(t) = u(t-1)$, $\theta_2^*(t) = -u(t-2)$, $\theta_3^*(t) = -u(t-3)$, $\theta_4^*(t) = u(t-4)$, and $f^*(t) = 0.5u(t-5)$. We list the four designs applied to the plant:

1. Basic design (Figure 2.2): $\omega_1 = \omega_2 = 5.48\,\text{rad/sec}$, $\omega_3 = \omega_4 = 6.32\,\text{rad/sec}$, $a_p = 0.05$, $\frac{C_{op}(s)}{\Gamma_f(s)} = \frac{s}{s+5}$, and $C_{ip}(s)\Gamma_{\theta p}(s) = 10\frac{1+0.01s}{s}$ for $p = 1,\ldots,4$. We found that, with this design, we could not achieve tracking with the initial conditions above, hence we show simulation results with plant variations an order of magnitude smaller, i.e., $\theta_1^*(t) = 0.1u(t-1)$, $\theta_2^*(t) = -0.1u(t-2)$, $\theta_3^*(t) = -0.1u(t-3)$, $\theta_4^*(t) = 0.1u(t-4)$, and $f^*(t) = 0.05u(t-5)$ in Figure 2.2.

2. High frequency design (Figure 2.3): $\omega_1 = \omega_2 = 5\,\text{rad/sec}$, $\omega_3 = \omega_4 = 14.14\,\text{rad/sec}$, $a_p = 0.1$, $\frac{C_{op}(s)}{\Gamma_f(s)} = \frac{s}{s+5}$, $C_{ip}(s)\Gamma_{\theta p}(s) = k_p\frac{1+0.01s}{s}$, for $p = 1,\ldots,4$ and $k_1 = k_2 = 30$, $k_3 = k_4 = 50$.

3. Low pass filter design (Figure 2.4): $\omega_1 = \omega_2 = 5.48\,\text{rad/sec}$, $\omega_3 = \omega_4 = 6.32\,\text{rad/sec}$, $a_p = 0.05$, $\frac{C_{op}(s)}{\Gamma_f(s)} = \frac{s}{s+5}$, $C_{ip}(s)\Gamma_{\theta p}(s) = 200\frac{1+0.01s}{s}F_{LPp}(s)$, $F_{LPp}(s) = \frac{s/z+1}{s/c+1}$, $z = 2.5$, and $c = \omega_3 - \omega_1 = 0.84$ for $p = 1,\ldots,4$.

4. Band pass filter design (Figure 2.5): $\omega_1 = \omega_2 = 5.48\,\text{rad/sec}$, $\omega_3 = \omega_4 = 6.32\,\text{rad/sec}$, $a_p = 0.05$, $\frac{C_{op}(s)}{\Gamma_f(s)} = \frac{s}{s+5}F_{BPp}(s)$, $F_{BPp}(s) = .0025\omega_p^2\frac{s^2+40s+1}{(s+5)^2+\omega_p^2}$, and $C_{ip}(s)\Gamma_{\theta p}(s) = 25\frac{1+0.01s}{s}$ for $p = 1,\ldots,4$.

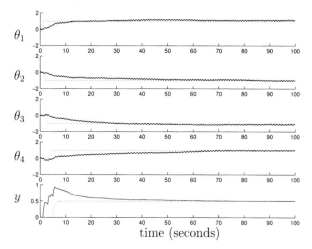

Figure 2.4: Step changes in plant: Low pass filter design

Comparing the four figures, we find that the three design variations shown in Figures 2.3, 2.4 and 2.5 perform equally well, and significantly better than the basic design in Figure 2.2 in tracking as well as output minimization, with initial conditions an order of magnitude larger.

2.3.2 General Variations in $\theta^*(t)$ and $f^*(t)$

The plant in this case has $F_{ip}(s) = 1$ for $p = 1, \ldots, 4$, $F_o(s) = \frac{2}{s+2}$, $f(\theta) = f^*(t) + (\theta - \theta^*(t))^T \mathbf{P}(\theta - \theta^*(t))$, $P_{pq} = 1$ if $p = q$, and $P_{pq} = 0.5$ otherwise. The plant variations are

$$\theta_1^*(t) = 0.01e^{0.01t},$$
$$\theta_2^*(t) = 0.2t,$$
$$\theta_3^*(t) = 0.1\sin t,$$
$$\theta_4^*(t) = 0.1u(t - 20),$$
$$f^*(t) = 0.01u(t - 30).$$

We list the four designs applied to the plant below.

1. Basic design (Figure 2.6): $\omega_1 = \omega_2 = 5\,\text{rad/sec}$, $\omega_3 = \omega_4 = 6.32\,\text{rad/sec}$, $a_p = .05$, $\frac{C_{op}(s)}{\Gamma_f(s)} = \frac{s}{s+5}$, $C_{i1}(s)\Gamma_{\theta 1}(s) = 10\frac{0.01s+5}{s-0.01}$, $C_{i2}(s)\Gamma_{\theta 2}(s) = 10\frac{s^2+6s+5}{s^2}$, $C_{i3}(s)\Gamma_{\theta 3}(s) = 10\frac{s^2+10.1s+1}{s^2+1}$, and $C_{i4}(s)\Gamma_{\theta 4}(s) = 10\frac{0.01s+10}{s}$ for $p = 1, \ldots, 4$. As in the case of step variations in plant parameters, we found that, with this design, we could not achieve tracking with the initial conditions above, hence we show simulation results with plant variations 10 times smaller, i.e., $\theta_1^*(t) = 0.001e^{0.01t}$, $\theta_2^*(t) = 0.02t$, $\theta_3^*(t) = 0.01\sin t$, $\theta_4^*(t) = 0.01u(t - 20)$, $f^*(t) = 0.001u(t - 30)$.

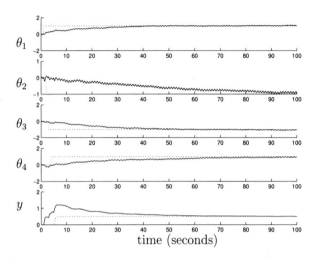

Figure 2.5: Step changes in plant: Band pass filter design

2. High frequency design (Figure 2.7): $\omega_1 = \omega_2 = 5\,\text{rad/sec}$, $\omega_3 = \omega_4 = 14.14\,\text{rad/sec}$, $a_p = .05$, $\frac{C_{op}(s)}{\Gamma_f(s)} = \frac{s}{s+5}$, and $C_{ip}(s)\Gamma_{\theta p}(s)$ the same as in design 1, for $p = 1, \ldots, 4$.

3. Low pass filter design (Figure 2.8): $\omega_1 = \omega_2 = 5\,\text{rad/sec}$, $\omega_3 = \omega_4 = 6.32\,\text{rad/sec}$, $a_p = 0.05$, $\frac{C_{op}(s)}{\Gamma_f(s)} = \frac{s}{s+5}$, $C_{ip}(s)\Gamma_{\theta p}(s) = C'_{ip}(s)\Gamma_{\theta p}(s)F_{LPp}(s)$,

$$C'_{i1}(s)\Gamma_{\theta 1}(s) = 10\frac{0.01s + 5}{s - 0.01}, \quad C'_{i2}(s)\Gamma_{\theta 2}(s) = 10\frac{s^2 + 6s + 5}{s^2},$$

$$C'_{i3}(s)\Gamma_{\theta 3}(s) = 15\frac{s^2 + 10.1s + 1}{s^2 + 1}, \quad \text{and} \quad C'_{i4}(s)\Gamma_{\theta 4}(s) = 15\frac{0.01s + 10}{s}$$

, $F_{LPp}(s) = \frac{s/z+1}{s/c+1}$, $z = 2.5$, and $c = \omega_3 - \omega_1 = 1.32$, for $p = 1, \ldots, 4$.

4. Band pass filter design (Figure 2.9): $\omega_1 = \omega_2 = 5\,\text{rad/sec}$, $\omega_3 = \omega_4 = 6.32\,\text{rad/sec}$, $a_p = 0.05$, $\frac{C_{op}(s)}{\Gamma_f(s)} = \frac{s}{s+5}F_{BPp}(s)$, $F_{BPp}(s) = .0025\omega_p^2\frac{s^2+40s+1}{(s+5)^2+\omega_p^2}$, $C'_{ip}(s)\Gamma_{\theta p}(s)$ the same as $C_{ip}(s)\Gamma_{\theta p}(s)$ in design 1, for $p = 1, \ldots, 4$.

A comparison of performance of the four designs again shows a significant performance improvement by the three design variations shown in Figures 2.7, 2.8, and 2.9 over the basic design in Figure 2.6.

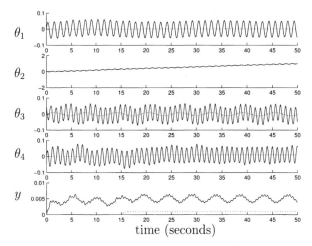

Figure 2.6: General plant variation: Basic design

Notes and References

Rotea [96] and Walsh [113] provided the first studies of multiparameter extremum seeking schemes. Their results were for plants with constant parameters. A systematic design procedure is absent in both [67, 70] and [96, 113]. This chapter is based upon [5] which supplied stability analysis for general multiparameter extremum seeking and systematic design guidelines for stability/performance. The need for Assumption 2.4 was recognized in [96].

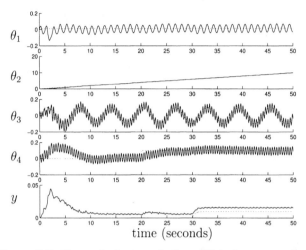

Figure 2.7: General plant variation: High frequency design

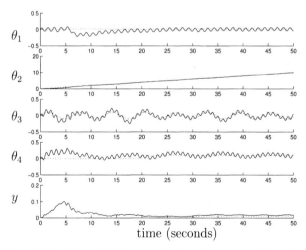

Figure 2.8: General plant variation: Low pass filter design

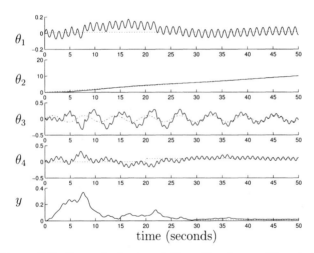

Figure 2.9: General plant variation: Band pass filter design

Chapter 3

Slope Seeking

Slope seeking is a recent idea for non-model based adaptive control introduced in [6]: *it involves driving the output of a plant to a value corresponding to a commanded slope of its reference-to-output map.* Extremum seeking is a special case of slope seeking where the commanded slope is zero. Motivations for the development of slope seeking are: problems where operation at the extremum of the plant reference-to-output map is susceptible to destabilization under finite disturbances, such as maximum pressure rise in deep hysteresis aeroengine compressors [116], antiskid braking for aircraft [110], minimum power demand formation flight [20], and problems in nuclear fusion where there is a need to stay away from the extremum (such as a maximal energy release condition) [55]. In all these problems, there is significant uncertainty in the models, and the set-points are unknown.

The results obtained herein constitute a generalization of perturbation-based extremum seeking, which seeks a point of zero slope, to the problem of seeking a general slope. With a small modification, the results on convergence in extremum seeking and the design guidelines derived in Chapter 1 and 2 are extended to permit system operation at a point of arbitrary slope on the reference-to-output map. The modification involves setting a reference slope in the algorithm, which, in extremum seeking, is implicitly set to zero. This chapter supplies the following results for enabling attainment of slope seeking feedback using sinusoidal perturbation:

1. Provides the problem formulation for the case of the plant being a simple static map (Section 3.1), the setting for classical extremum seeking schemes. It next supplies the formulation of single parameter slope seeking for the general case where the map is embedded within dynamics with time-varying parameters (Section 3.2).

2. Presents derivation of a stability test for single parameter slope seeking (Section 3.2) and systematic design guidelines using standard linear SISO control design methods to satisfy the stability test (Section 3.3).

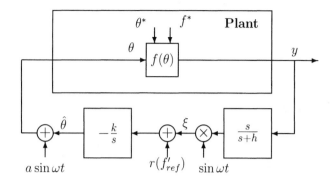

Figure 3.1: Basic slope seeking scheme

3. Supplies the extension of the above results to the multivariable case of gradient seeking (Section 3.4).

For ease of understanding of the method, we present the result with slope seeking on a static map in Section 3.1 accompanied by an illustrative simulation. Section 3.2 presents the analysis, and Section 3.3 the design algorithm for generalized single parameter slope seeking; Section 3.4 supplies results on multiparameter gradient seeking.

3.1 Slope Seeking on a Static Map

Figure 3.1 shows a basic slope seeking loop for a static map. We posit $f(\theta)$ of the form:

$$f(\theta) = f^* + f'_{ref}(\theta - \theta^*) + \frac{f''}{2}(\theta - \theta^*)^2 \qquad (3.1)$$

where f'_{ref} is the *commanded slope we want to operate at*, and $f'' > 0$. Any C^2 function $f(\theta)$ can be approximated locally by Eqn. (3.1). The assumption $f'' > 0$ is made without loss of generality. If $f'' < 0$, we just replace k $(k > 0)$ in Figure 3.1 with $-k$. The purpose of the algorithm is to make $\theta - \theta^*$ as small as possible, so that the output $f(\theta)$ is driven to its optimum f^*.

The perturbation signal $a \sin \omega t$ into the plant helps to give a measure of gradient information of the map $f(\theta)$. This is obtained by removing f^* from the output using the washout filter $\frac{s}{s+h}$ $(h > 0)$, and then demodulating the signal with $\sin \omega t$. In a sense, this can also be thought of as the online extraction of a Fourier coefficient. The input $r(f'_{ref})$ serves as a slope set point which is explicitly calculated below.

Output Optimization. The following bare-bones result sums up the properties of the rudimentary slope seeking loop in Figure 3.1:

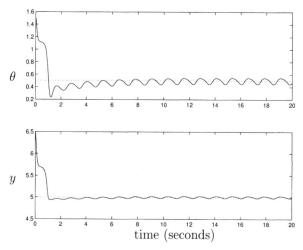

Figure 3.2: Slope seeking, $r(f'_{ref}) = -0.00625$

Theorem 3.1 (Slope Seeking) *For the system in Figure 3.1, the output error $y - f^*$ achieves local exponential convergence to an $O(a + 1/\omega)$ neighborhood of the origin provided the perturbation frequency ω is sufficiently large, $\frac{1}{1 + L(s)}$ is asymptotically stable, where*

$$L(s) = \frac{kaf''}{2s}, \qquad (3.2)$$

and provided

$$r(f'_{ref}) = -\frac{af'_{ref}}{2} \operatorname{Re}\left\{ \frac{j\omega}{j\omega + h} \right\}. \qquad (3.3)$$

We omit the proof as this result is subsumed in a more general result we prove in the following section. The result in Theorem 3.1 has the following salient features:

1. Like the analogous result on extremum seeking in Chapter 1, it provides a linear stability test permitting design using linear SISO control tools.

2. Provided we know the sign of the second derivative f'' in the neighborhood, we can create a feedback that drives the system to operate at a prespecified slope f'_{ref} of the input-output map; this is done exactly through setting the reference $r(f'_{ref}) = -\frac{af'_{ref}}{2} \operatorname{Re}\left\{ \frac{j\omega}{j\omega + h} \right\}$.

3. Unlike the extremum seeking result, the convergence is only first order, i.e., $O(a + 1/\omega)$; this is because we are seeking a point of non-zero slope.

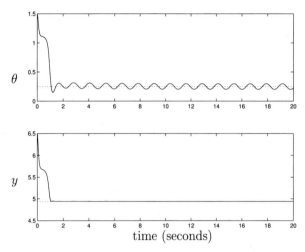

Figure 3.3: Extremum seeking, $r(f'_{ref}) = 0$

Simulation Example. We present an example to illustrate the method proposed above. Simulation results are plotted with $f^*(t), \theta^*(t)$ in dotted lines and y, θ in solid lines. We use the static map $f(\theta) = f^* + 0.5(\theta - \theta^*) + (\theta - \theta^*)^2$, where $f^*(t) = 5.0$, and $\theta^* = 0.5$.

To satisfy the conditions in Theorem 3.1, we set $\omega = 5$ rad/sec, $a = 0.05$, washout filter $\frac{s}{s+h}$ with $h = 5.0$, integrator gain $k = 10$, and slope setting $r(f'_{ref}) = -\frac{af'_{ref}}{2} \text{Re}\left\{\frac{j5}{j5+5}\right\} = -0.00625$ for operating at the slope $f'_{ref} = 0.5$. Substituting all parameters in Eqn. (3.2) we get

$$L(s) = \frac{1}{4s}, \tag{3.4}$$

and attain stable slope seeking (Figure 3.2) through stability of $\frac{1}{1+L(s)}$. An extremum seeking design for the same plant with $r(f'_{ref}) = 0$, and other design parameters the same as for slope seeking, is shown in Figure 3.3 for comparison. Extremum seeking tracks a slope set point of zero, the minimum at $\theta = 0.25$ of the map $f(\theta)$ ($f(0.25) = 4.9375 < f(0.5) = 5$).

3.2   General Single Parameter Slope Seeking

The generalized scheme differs from the rudimentary scheme of Figure 3.1 in the following ways: the map has time varying parameters and is embedded amidst linear dynamics; the slope seeking loop incorporates parameter dynamics for tracking parameter variations. Figure 3.4 shows the time-varying nonlinear map embedded amidst linear dynamics along with the slope seeking

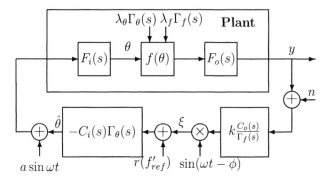

Figure 3.4: Generalized slope seeking

loop. We posit $f(\theta)$ with time-varying parameters of the form:

$$f(\theta) = f^*(t) + f'_{ref}(\theta - \theta^*(t)) + \frac{f''}{2}(\theta - \theta^*(t))^2 \tag{3.5}$$

where $f'' > 0$, and f'_{ref} is the commanded slope. Any C^2 function $f(\theta)$ can be approximated locally by Eqn. (3.5). The assumption $f'' > 0$ is made without loss of generality. If $f'' < 0$, we just replace $C_i(s)$ in Figure 3.4 with $-C_i(s)$. The purpose of the algorithm is to make $\theta - \theta^*$ as small as possible, so that the output $F_o(s)[f(\theta)]$ is driven to its optimum $F_o(s)[f^*(t)]$. As in Chapters 1, 2, n denotes measurement noise. Before proceeding to the analysis, we make the following assumptions:

Assumption 3.2 $F_i(s)$ and $F_o(s)$ are asymptotically stable and proper.

Assumption 3.3 $\mathcal{L}\{f^*(t)\} = \lambda_f\Gamma_f(s)$ and $\mathcal{L}\{\theta^*(t)\} = \lambda_\theta\Gamma_\theta(s)$ are strictly proper rational functions and poles of $\Gamma_\theta(s)$ that are not asymptotically stable are not zeros of $F_i(s)$.

This assumption forbids delta function variations in the map parameters and also the situation where tracking of the extremum is not possible.

Assumption 3.4 $\frac{C_o(s)}{\Gamma_f(s)}$ and $C_i(s)\Gamma_\theta(s)$ are proper.

This assumption ensures that the filters $\frac{C_o(s)}{\Gamma_f(s)}$ and $C_i(s)\Gamma_\theta(s)$ in Figure 3.4 can be implemented. Since $C_i(s)$ and $C_o(s)$ are at our disposal to design, we can always satisfy this assumption.

Although LTV stability tests can be given for general slope seeking loops analogous to the results in Propositions 1.5, 2.5 simply by introducing a slope setting, we do not provide them here in the interest of compactness. We provide only the results through which systematic design is possible. To this end, we introduce the following notation:

$$H_o(s) = k\frac{C_o(s)}{\Gamma_f(s)}F_o(s) \triangleq H_{osp}(s)H_{obp}(s) \triangleq H_{osp}(s)(1 + H^{sp}_{obp}(s)) \quad (3.6)$$

where $H_{osp}(s)$ denotes the strictly proper part of $H_o(s)$ and $H_{obp}(s)$ its biproper part, k is chosen to set

$$\lim_{s \to 0} H_{osp}(s) = 1, . \quad (3.7)$$

The two assumptions below are analogous to Assumptions 1.6, 1.7:

Assumption 3.5 *Let the smallest in absolute value among the real parts of all of the poles of $H_{osp}(s)$ be denoted by a. Let the largest among the moduli of all of the poles of $F_i(s)$ and $H_{obp}(s)$ be denoted by b. The ratio $M = a/b$ is sufficiently large.*

Assumption 3.6 *$H_i(s)$ is strictly proper.*

Output Optimization. We first provide background for the result on slope seeking below. The following equations describe the single parameter slope seeking scheme in Fig. 3.4:

$$y = F_o(s)\left[f^*(t) + f'_{ref}(\theta - \theta^*(t)) + \frac{f''}{2}(\theta - \theta^*(t))^2\right] \quad (3.8)$$

$$\theta = F_i(s)\left[a\sin(\omega t) - C_i(s)\Gamma_\theta(s)[\xi + r(f'_{ref})]\right] \quad (3.9)$$

$$\xi = \sin(\omega t - \phi)k\frac{C_o(s)}{\Gamma_f(s)}[y + n]. \quad (3.10)$$

For the purpose of analysis, we define the tracking error $\tilde{\theta}$ and output error $\tilde{y}$:

$$\tilde{\theta} = \theta^*(t) - \theta + \theta_0 \quad (3.11)$$

$$\theta_0 = F_i(s)[a\sin(\omega t)] \quad (3.12)$$

$$\tilde{y} = y - F_o(s)[f^*(t)] \quad (3.13)$$

In terms of these definitions, we can restate the goal of slope seeking as driving output error $\tilde{y}$ to a small value by tracking $\theta^*(t)$ with θ. With the present method, we cannot drive $\tilde{y}$ to zero because of the sinusoidal perturbation θ_0. We are now ready for our single parameter result:

Theorem 3.7 (Slope Seeking) *For the system in Figure 3.4, under Assumptions 3.2–3.6, the output error $\tilde{y}$ achieves local exponential convergence to an $O(a + \delta)$ neighborhood of the origin, where $\delta = 1/\omega + 1/M$, provided $n = 0$ and:*

1. *Perturbation frequency ω is sufficiently large, and $\pm j\omega$ is not a zero of $F_i(s)$.*

2. *Zeros of $\Gamma_f(s)$ that are not asymptotically stable are also zeros of $C_o(s)$.*

3. *Poles of $\Gamma_\theta(s)$ that are not asymptotically stable are not zeros of $C_i(s)$.*

4. *$C_o(s)$ and $\frac{1}{1+L(s)}$ are asymptotically stable, where*

$$L(s) = \frac{af''}{4}\operatorname{Re}\{e^{j\phi}F_i(j\omega)\}H_i(s), \tag{3.14}$$

$$H_i(s) = C_i(s)\Gamma_\theta(s)F_i(s). \tag{3.15}$$

5. $r(f'_{ref}) = -\frac{af'_{ref}}{2}\mathbf{Re}\{e^{-j\phi}H_o(j\omega)F_i(j\omega)\}.$

Proof: Using $n = 0$ and substituting Eqns. (3.9) and (3.12) in Eqn. (3.11) yields

$$\tilde{\theta} = \theta^* + H_i(s)[\xi + r(f'_{ref})] \tag{3.16}$$

Further, substitution for ξ from Eqn. (3.10) and for y from Eqn. (3.8) yields

$$\tilde{\theta} = \theta^* + H_i(s)\left[\sin(\omega t - \phi)H_o(s)\left[f^* + f'_{ref}(\theta - \theta^*) + \frac{f''}{2}(\theta - \theta^*)^2\right] + r(f'_{ref})\right] \tag{3.17}$$

Using $\theta - \theta^* = \theta_0 - \tilde{\theta}$ from Eqn. (3.11), we get

$$\tilde{\theta} = \theta^* + H_i(s)\left[\sin(\omega t - \phi)H_o(s)\left[f^* + f'_{ref}(\theta_0 - \tilde{\theta}) + \frac{f''}{2}(\theta_0 - \tilde{\theta})^2\right] + r(f'_{ref})\right]$$

$$= \theta^* + H_i(s)\left[\sin(\omega t - \phi)H_o(s)\left[f^* + f'_{ref}\theta_0 - f'_{ref}\tilde{\theta} + \frac{f''}{2}(\theta_0^2 - 2\theta_0\tilde{\theta} + \tilde{\theta}^2)\right]\right.$$

$$\left. + r(f'_{ref})\right] \tag{3.18}$$

We drop the higher order term[1] $\tilde{\theta}^2$ and simplify the expression in Eqn. (3.18) using Lemmas A.1, A.2, Assns. 3.2, 3.3, and 3.4 and asymptotic stability of $\frac{C_o(s)}{\Gamma_f(s)}$ and $C_o(s)$:

$$\sin(\omega t - \phi)H_o(s)[f^*(t)] = \lambda_f \sin(\omega t - \phi)\mathcal{L}^{-1}\left(H_o(s)\Gamma_f(s)\right)$$

$$= \sin(\omega t - \phi)(\epsilon^{-t}) = \epsilon^{-t} \tag{3.19}$$

$$\sin(\omega t - \phi)H_o(s)[\theta_0^2] = C_1a^2\sin(\omega t + \mu_1) + C_2a^2\sin(3\omega t + \mu_2) + \epsilon^{-t} \tag{3.20}$$

$$\sin(\omega t - \phi)H_o(s)[f'_{ref}\theta_0] = \frac{af'_{ref}}{2}\left(\mathbf{Re}\{e^{-j\phi}H_o(j\omega)F_i(j\omega)\}\right.$$

$$\left. - \mathbf{Re}\{e^{j(2\omega t - \phi)}H_o(j\omega)F_i(j\omega)\}\right) + \epsilon^{-t}, \tag{3.21}$$

---

[1]As in the proof of Proposition 1.5, this is justified by Lyapunov's first method, as we have already written the system in terms of error variable $\tilde{\theta}$ thus transforming the problem to stability of the origin. As in the proof of Theorem 1.8, this is responsible for the result in the theorem being local.

where C_1, C_2, μ_1, μ_2 are constants (these can be determined from the frequency response of $H_o(s)$), and ϵ^{-t} denotes exponentially decaying terms. Hence, after substituting Eqns. (3.19), (3.20), (3.21) in Eqn. (3.18) we can write the linearization of Eqn. (3.18) as

$$\tilde{\theta} \;=\; \theta^* + H_i(s)\left[\sin(\omega t - \phi)H_o(s)\left[-f'_{ref}\tilde{\theta} - f''\theta_0\tilde{\theta}\right] + w(t) + \epsilon^{-t}\right] \quad (3.22)$$

$$w(t) \;=\; a^2\frac{f''}{2}\left[C_1\sin(\omega t + \mu_1) + C_2\sin(3\omega t + \mu_2)\right]$$

$$+\frac{af'_{ref}}{2}\mathbf{Re}\{e^{j(2\omega t-\phi)}H_o(j\omega)F_i(j\omega)\}, \quad (3.23)$$

where we have used

$$r(f'_{ref}) = -\frac{af'_{ref}}{2}\mathbf{Re}\{e^{-j\phi}H_o(j\omega)F_i(j\omega)\}.$$

Applying the reduction of $H_o(s)$ from Assumption 3.5 and Lemmas A.1, A.2 in succession to the terms containing $2\theta_0\tilde{\theta}$ and $f'_{ref}\tilde{\theta}$ in Eqn. (3.22), we get[2]

$$H_i(s)\left[\sin(\omega t - \phi)H_o(s)[-f''\theta_0\tilde{\theta} - f'_{ref}\tilde{\theta}]\right]$$

$$= H_i(s)\left[\sin(\omega t - \phi)(1 + H^{sp}_{obp}(s))[-f''\theta_0\tilde{\theta} + f'_{ref}\tilde{\theta}]\right] \quad (3.24)$$

$$= \mathcal{T}[\tilde{\theta}] - L(s)[\tilde{\theta}] - \mathcal{S}[\tilde{\theta}] + H_i(s)[\sin(\omega t - \phi)v_0(t)], \quad (3.25)$$

where

$$L(s)[\tilde{\theta}] \;=\; \frac{af''}{2}H_i(s)\left[\mathrm{Re}\left\{e^{j\phi}F_i(j\omega)[\tilde{\theta}]\right\}\right] \quad (3.26)$$

$$\mathcal{T}[\tilde{\theta}] \;=\; \frac{af''}{2}H_i(s)\left[\mathrm{Re}\left\{e^{j(2\omega t-\phi)}F_i(j\omega)[\tilde{\theta}]\right\}\right] \quad (3.27)$$

$$\mathcal{S}[\tilde{\theta}] \;=\; f'_{ref}H_i(s)[\sin(\omega t - \phi)\tilde{\theta}] \quad (3.28)$$

$$v_0(t) \;=\; H^{sp}_{obp}(s)\left[-f''\mathrm{Im}\{aF_i(j\omega)e^{j\omega t}\}\tilde{\theta} + f'_{ref}\tilde{\theta}\right]. \quad (3.29)$$

Finally, substtituting Eqn. (3.25) in Eqn. (3.22), and moving the terms linear in $\tilde{\theta}$ to the left hand side, we get

$$(1 + L(s) - \mathcal{T} + \mathcal{S})[\tilde{\theta}] - H_i(s)[\sin(\omega t - \phi)v_0(t)]$$

$$= \theta^* + H_i(s)\left[w(t) + \epsilon^{-t}\right]. \quad (3.30)$$

---

[2]Note that Eqn. (3.25) contains an additional term of the form $H_i(s)[\sin(\omega t - \phi)H_o(s)[\epsilon^{-t}\tilde{\theta}]]$ which comes from ϵ^{-t} in $\theta_0(t) = a\mathrm{Im}\{F_i(j\omega)e^{j\omega t}\} + \epsilon^{-t}$. We drop this term from subsequent analysis because it does not affect closed loop stability or asymptotic performance. It can be accounted for in three ways. One is to perform averaging over an infinite time interval in which all exponentially decaying terms disappear. The second way is to treat $\epsilon^{-t}\tilde{\theta}$ as a vanishing perturbation via Corollary 5.4 in Khalil [64], observing that ϵ^{-t} is integrable. The third way is to express ϵ^{-t} in state space format and let $\epsilon^{-t}\tilde{\theta}$ be dominated by other terms in a local Lyapunov analysis.

We now divide both sides of Eqn. (3.30) with $1 + L(s)$ and rewrite it as

$$\tilde{\theta} - Y_i(s) \left[af''/2\mathrm{Re}\{e^{j(2\omega t - \phi)}\tilde{\theta}\} + af'_{ref}\sin(\omega t - \phi)\tilde{\theta} + \sin(\omega t - \phi)v_0(t) \right]$$
$$= \frac{1}{1 + L(s)}[\theta^*] + Y_i(s)\left[w(t) + \epsilon^{-t}\right], \tag{3.31}$$

where $Y_i(s) = \frac{H_i(s)}{1+L(s)} = \frac{\mathrm{num}\{Y_i(s)\}}{\mathrm{num}\{1+L(s)\}}$ is asymptotically stable because the poles of $H_i(s)$ are cancelled by zeros of $\frac{1}{1+L(s)}$, and $\frac{1}{1+L(s)}$ is asymptotically stable. By noting also that zeros in $\frac{1}{1+L(s)}$ cancel poles in $\theta^*(s) = \lambda_\theta \Gamma_\theta(s)$, and using asymptotic stability of $\frac{1}{1+L(s)}$, we get

$$\tilde{\theta} - Y_i(s)\left[af''/2\mathrm{Re}\{e^{j(2\omega t - \phi)}\tilde{\theta}\} + af'_{ref}\sin(\omega t - \phi)\tilde{\theta} + \sin(\omega t - \phi)v_0(t)\right]$$
$$= \epsilon^{-t} + Y_i(s)\left[w(t)\right] \tag{3.32}$$

Now, $Y_i(s)$ is strictly proper, and can therefore be written as $Y_i(s) = \frac{1}{s+p_0}Y_i'(s)$, where $Y_i'(s)$ is proper. In terms of their partial fraction expansions, we can write $Y_i'(s) = A_0 + \sum_{k=1}^{n}\frac{A_k}{s+p_k}$, and $H_{obp}^{sp}(s) = \sum_{j=1}^{m}\frac{B_j}{s+p_j}$. Multiplying both sides of Eqn. (3.32) with $s + p_0$ and using the partial fraction expansions, we get

$$\dot{\tilde{\theta}} + p_0\tilde{\theta} - A_0(u_0(t) + \sin(\omega t - \phi)v_0(t) - w(t))$$
$$- \sum_{k=1}^{n}(u_k(t) + v_k(t) - w_k(t)) = \epsilon^{-t} \tag{3.33}$$

$$u_0(t) = af''/2\mathrm{Re}\{e^{j(2\omega t - \phi)}\tilde{\theta}\} + af'_{ref}\sin(\omega t - \phi)\tilde{\theta}$$

$$u_k(t) = \frac{A_k}{s+p_k}[u_0(t)], v_k(t) = \frac{A_k}{s+p_k}[\sin(\omega t - \phi)v_0(t)]$$

$$w_k(t) = \frac{A_k}{s+p_k}[w(t)]$$

$$v_0(t) = \sum_{j=1}^{m}v_{1j}(t), v_{1j}(t) = \frac{B_j}{s+p_j}\left[-f''\mathrm{Im}\{aF_i(j\omega)e^{j\omega t}\}\tilde{\theta} + f'_{ref}\tilde{\theta}\right]. \tag{3.34}$$

We can write the system of linear time varying differential equations above in the state-space form (as was done in Chapter 1):

$$\dot{\mathbf{x}} = A(t)\mathbf{x} + A_{12}\mathbf{x}_e + Bw(t); \quad \tilde{\theta} = x_1 \tag{3.35}$$
$$\dot{\mathbf{x}}_e = A_e\mathbf{x}_e. \tag{3.36}$$

Eqn. (3.36) is a representation for the ϵ^{-t}. We get Eqns. (3.35), (3.36) into the standard form for averaging by using the transformation $\tau = \omega t$, and then averaging the right hand side of the equations w.r.t. time from 0 to $T = 2\pi/\omega$, i.e., $\frac{1}{T}\int_0^T (\cdot)d\tau$ treating states $\mathbf{x}$, $\mathbf{x}_e$ as constant to get:

$$\frac{d\mathbf{x}_{av}}{d\tau} = \frac{1}{\omega}\left(A_{av}\mathbf{x}_{av} + A_{12}\mathbf{x}_{eav}\right),\ \tilde{\theta}_{av} = x_{1av} \qquad (3.37)$$

$$\frac{d\mathbf{x}_{eav}}{d\tau} = \frac{1}{\omega}A_e\mathbf{x}_{eav}, \qquad (3.38)$$

which is a state-space representation of the system in the $\tau = \omega t$ time-scale, and $A_{av} = \frac{1}{T}\int_0^T (A(\tau))d\tau$. This gives

$$\dot{\tilde{\theta}}_{av} + p_0\tilde{\theta}_{av} - \sum_{k=1}^{n}(u_{k,av} + v_{k,av} - w_{k,av}) = \epsilon^{-t} \qquad (3.39)$$

$$\dot{u}_{k,av} + p_k u_{k,av} = 0,\ \dot{v}_{k,av} + p_k v_{k,av} = 0,\ \dot{w}_{k,av} + p_k w_{k,av} = 0$$

$$v_{0,av} = \sum_{j=1}^{m} v_{1j,av},\ \dot{v}_{1j,av} + p_j v_{1j,av} = B_j f'_{ref}\tilde{\theta}_{av}. \qquad (3.40)$$

in the original time-scale. As all of the poles p_k for all k and p_j for all j are asymptotically stable (from asymptotic stability of $H_o(s)$ and $\frac{1}{1+L(s)}$), all of the terms on the right hand side of Eqn. (3.39) for $\tilde{\theta}_{av}$ are exponentially decaying, i.e., we have

$$\dot{\tilde{\theta}}_{av} + p_0\tilde{\theta}_{av} = \epsilon^{-t}, \qquad (3.41)$$

which decays to zero because p_0, a pole of $\frac{1}{1+L(s)}$ is asymptotically stable. Hence, by a standard averaging theorem such as Theorem 8.3 in Khalil [64], we see that if ω, a, ϕ, $C_i(s)$ and $C_o(s)$ are such that $\frac{1}{1+L(s)}$ is asymptotically stable, and ω is sufficiently large relative to other parameters of the state-space representation, solutions starting from small initial conditions converge exponentially to a periodic solution in an $O(1/\omega)$ neighborhood of zero. Hence, $\tilde{\theta}(t)$ goes to a periodic solution $\tilde{\theta}_{per}(t) = O(1/\omega)$. We now proceed to put the system in the standard form for singular perturbation analysis through making the transformation $\delta\tilde{\theta} = \tilde{\theta}(t) - \tilde{\theta}_{per}(t)$ in the unreduced linearized system in Eqn. (3.22) and get:

$$\delta\tilde{\theta} + \tilde{\theta}_{per}(t) = \theta^* + H_i(s)\left[\sin(\omega t - \phi)[y'_{osp}] + w(t) + \epsilon^{-t}\right] \qquad (3.42)$$

$$y'_{osp} = (1 + H^{sp}_{obp}(s))[y_{osp}]$$

$$y_{osp} = H_{osp}(s)\left[(f''\theta_0 - f'_{ref})(\delta\tilde{\theta} + \tilde{\theta}_{per})\right] \qquad (3.43)$$

By linearity of the system described by the Eqns. (3.42), (3.43), we have that the reduced order model in the new coordinates (replacing $H_{osp}(s)$ with its unity static gain) is given by

$$\delta\tilde{\theta} = H_i(s)\left[\sin(\omega t - \phi)[y'_{osp}]\right]$$

$$y'_{osp} = -(1 + H^{sp}_{obp}(s))\left[(f''\theta_0 + f'_{ref})\delta\tilde{\theta}\right], \qquad (3.44)$$

which has the state space representation

$$\dot{\mathbf{x}} = A(t)\mathbf{x}; \quad \delta\tilde{\theta} = x_1, \tag{3.45}$$

where $A(t)$ is the same as in Eqn.(3.35). Hence $\delta\tilde{\theta}$ converges exponentially to the origin. This shows that the reduced order model is exponentially stable. From exponential stability of $H_{osp}(s)$, we have exponential stability of the boundary layer model

$$\frac{d\mathbf{y}}{d\tau} = \mathbf{A}_{osp}\mathbf{y}, \tag{3.46}$$

where $(\mathbf{A}_{osp}, \mathbf{B}_{osp}, \mathbf{C}_{osp})$ is a state space representation of $H_{osp}(s)$, with $\mathbf{C}_{osp}\mathbf{A}_{osp}^{-1}\mathbf{B}_{osp} = 1$ from Eqn. (3.6). Hence, by the Singular Perturbation Lemma A.3, we have that in the overall unreduced system in Eqns. (3.42), (3.43), the solution converges to an O$(1/M)$ neighborhood of the origin. Hence, $\delta\tilde{\theta}(t)$ converges to a O$(1/M)$ neighborhood of the origin. Therefore, $\tilde{\theta}$ converges exponentially to a O$(1/\omega)$ + O$(1/M)$ = O(δ) neighborhood of the origin. Further, the output error $\tilde{y}$ decays to O$(a + \delta)$:

$$
\begin{aligned}
\tilde{y} &= F_o(s)\left[f'_{ref}(\theta - \theta^*) + \frac{f''}{2}(\theta - \theta^*)^2\right] \\
&= F_o(s)\left[f'_{ref}(\tilde{\theta} - \theta_0) + \frac{f''}{2}(\tilde{\theta} - \theta_0)^2\right] = \text{O}(a + \delta), \tag{3.47}
\end{aligned}
$$

dropping second order terms, which completes the proof.   Q.E.D.

The output error $\tilde{y}$ converges to an O$(a + 1/\omega)$ neighborhood of the origin. Thus, the deviation of the output from the desired output will be larger than that achievable in extremum seeking, where we track a point on the map with zero first derivative. We next provide rigorous design guidelines that satisfy the conditions of Theorem 3.7. We now note that *for $r(f'_{ref}) = 0$, the slope seeking scheme reduces to the extremum seeking scheme in Chapter 1.*

3.3 Compensator Design

In the design guidelines that follow, we set $\phi = 0$ which can be used separately for fine-tuning.

Algorithm 3.3.1 (Single Parameter Slope Seeking)

1. *Select the perturbation frequency ω sufficiently large. Also, ω should not equal any frequency in noise.*

2. *Set perturbation amplitude a so as to obtain small steady state output error $\tilde{y}$.*

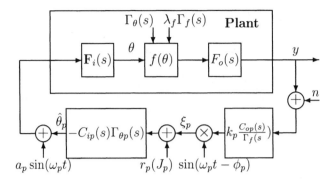

Figure 3.5: Multiparameter gradient seeking with $p = 1, 2, \ldots, l$.

3. *Design $C_o(s)$ asymptotically stable, with zeros of $\Gamma_f(s)$ that are not asymptotically stable as its zeros, and such that $\frac{C_o(s)}{\Gamma_f(s)}$ is proper. In the case where dynamics in $F_o(s)$ are slow and strictly proper, use as many fast fast poles in $C_o(s)$ as the relative degree of $F_o(s)$, and as many zeros as needed to have zero relative degree of the slow part $H_{obp}(s)$ to satisfy Assumption 3.5.*

4. *Design $C_i(s)$ by any linear SISO design technique such that it does not include poles of $\Gamma_\theta(s)$ that are not asymptotically stable as its zeros, $C_i(s)\Gamma_\theta(s)$ is proper, and $\frac{1}{1+L(s)}$ is asymptotically stable.*

5. *Set $r(f'_{ref}) = -\frac{af'_{ref}}{2}\mathbf{Re}\{e^{-j\phi}H_o(j\omega)F_i(j\omega)\}$.*

Steps 1,..., 4 are discussed fully in Chapter 1. The setting of $r(f'_{ref})$ requires knowledge of the frequency response of $F_i(s)$ and $F_o(s)$ at ω. We note here that seeking large slopes is difficult because $\tilde{y}$ will be correspondingly large from Eqn. (3.47).

Convergence of the scheme requires asymptotic stability of $\frac{1}{1+L(s)}$, and this requires knowledge of the second derivative f'' of the map at θ^*, or robustness to a range of values of f''. This is dealt with in Chapter 1.

3.4   Multiparameter Gradient Seeking

The results on multiparameter extremum seeking presented in Chapter 2 can be extended to gradient seeking through setting reference inputs in each of the parameter tracking loops. Figure 3.5 shows the multiparameter gradient seeking scheme with reference inputs r_p ($r_p = 0$ in each loop corresponds to the multiparameter extremum seeking scheme in Chapter 2). Analogous to the single parameter case in Section 3.2, we let $f(\theta)$ be a function of the form:

$$f(\theta) = f^*(t) + \mathbf{J}^T(\theta - \theta^*(t)) + (\theta - \theta^*(t))^T \mathbf{P}(\theta - \theta^*(t)), \qquad (3.48)$$

where $\mathbf{P}_{l \times l} = \mathbf{P}^T$, $\theta = [\theta_1 \ldots \theta_l]^T$, $\theta^*(t) = [\theta_1^*(t) \ldots \theta_l^*(t)]^T$, $\mathcal{L}\{\theta^*(t)\} = \Gamma_\theta(s)$ $= [\lambda_1 \Gamma_{\theta 1}(s), \ldots, \lambda_l \Gamma_{\theta l}(s)]^T$, $\mathcal{L}\{f^*(t)\} = \lambda_f \Gamma_f(s)$, and $\mathbf{J} = [J_1, J_2, \ldots, J_l]$ is the *commanded gradient*. Any twice differentiable vector function $f(\theta)$ can be approximated by Eqn. (3.48). As in multiparameter extremum seeking, the broad principle of using m frequencies for identification/tracking of $2m$ parameters applies; but for simplicity of presentation, we only present the case where a separate forcing frequency is used in each parameter tracking loop, i.e., we use forcing frequencies $\omega_1 < \omega_3 < \ldots < \omega_l$. We make assumptions identical to those made to prove Theorem 2.8:

Assumption 3.8 $\mathbf{F}_i(s) = [F_{i1}(s) \ldots F_{il}(s)]^T$ *and* $F_o(s)$ *are asymptotically stable and proper.*

Assumption 3.9 $\Gamma_\theta(s)$ *and* $\Gamma_f(s)$ *are strictly proper.*

Assumption 3.10 $C_{ip}(s) \Gamma_{\theta p}(s)$ *and* $\frac{C_{op}(s)}{\Gamma_f(s)}$ *are proper for all* $p = 1, 2, \ldots, l$.

Assumption 3.11 $\omega_p + \omega_q \neq \omega_r$ *for any* $p, q, r = 1, 2, \ldots, l$.

As for multiparameter extremum seeking in Chapter 2, we introduce the following notation for the next assumption:

$$H_{op}(s) = k_p \frac{C_{op}(s)}{\Gamma_f(s)} F_o(s) \triangleq H_{osp,p}(s) H_{obp,p}(s)$$

$$\triangleq H_{osp,p}(s)(1 + H_{obp,p}^{sp}(s)) \tag{3.49}$$

$$\lim_{s \to 0} H_{osp,p}(s) = 1,$$

where $H_{osp,p}(s)$ denotes the strictly proper part of $H_{op}(s)$ and $H_{obp,p}(s)$ its biproper part, $k_p, p = 1, \ldots, l$ is chosen to normalize the static gain of $H_{osp,p}(s)$ to unity.

Assumption 3.12 *Let the smallest in absolute value among the real parts of all of the poles of* $H_{osp,p}(s)$ *for all* p *be denoted by* a. *Let the largest among the moduli of all of the poles of* $F_{ip}(s)$ *and* $H_{obp,p}(s)$ *for all* p, *be denoted by* b. *The ratio* $M = a/b$ *is sufficiently large.*

Theorem 3.13 (Multiparameter Gradient Seeking) *For the system in Figure 3.5, under Assumptions 3.8–3.12, the output* y *achieves local exponential convergence to an* $O(\sum_{p=1}^l a_p + \Delta)$ *neighborhood of* $F_o(s)[f^*(t)]$ *provided* $n = 0$ *and:*

1. *Perturbation frequencies* $\omega_1 < \omega_2 < \ldots < \omega_l$ *are rational, sufficiently large, and* $\pm j\omega_p$ *is not a zero of* $F_{ip}(s)$.

2. *Zeros of* $\Gamma_f(s)$ *that are not asymptotically stable are also zeros of* $C_{op}(s)$, *for all* $p = 1, \ldots, l$.

3. *Poles of $\Gamma_{\theta p}(s)$ that are not asymptotically stable are not zeros of $C_{ip}(s)$, for any $p = 1, \ldots, l$.*

4. *$C_{op}(s)$ are asymptotically stable for all $p = 1, \ldots, l$ and $\frac{1}{\det(\mathbf{I}_l + \mathbf{X}(s))}$ is asymptotically stable, where $X_{pq}(s)$ denote the elements of $\mathbf{X}(s)$ and*

$$X_{pq}(s) = P_{pq}a_p L_p(s), \quad q = 1, \ldots, l \qquad (3.50)$$

$$L_p(s) = \frac{1}{4} H_{ip}(s) \operatorname{Re}\{e^{j\phi_p} F_{ip}(j\omega_p)\}, \qquad (3.51)$$

where $H_{ip}(s) = C_{ip}(s)\Gamma_{\theta p}(s)F_{ip}(s)$ and $\Delta = 1/\omega + 1/M$.

5. *The reference is chosen as*

$$r_p(J_p) = -\frac{a_p J_p}{2} \operatorname{\mathbf{Re}}\{e^{-j\phi_p} H_{op}(j\omega_p)F_{ip}(j\omega_p)\}, \quad p = 1, \ldots, l$$

The proof is a simple extension of the proof of Theorem 2.8. Additional terms produced by the gradient term in Eqn. (3.48) are handled without any difficulty by the method of averaging. The key point to note in this result is that *the greater the number of parameters, the poorer the convergence.* Furthermore, the design guidelines in 2.2 apply to gradient seeking with the added specification of the components of the gradient, $r_1(J_1), r_2(J_2), \ldots, r_l(J_l)$ by Theorem 3.13.

Notes and References

In work on aircraft antiskid control, Tunay [110] used a slope set point in an extremum seeking loop to ameliorate the problem of system instability at the point of maximal friction. This chapter is based upon the results in [6] which developed analysis based design of slope seeking. The results obtained therein constitute a generalization of perturbation-based extremum seeking, which seeks a point of zero slope, to the problem of seeking a general slope. With a small modification, the results on convergence in extremum seeking and the design guidelines derived therefrom [5] were extended to permit system operation at a point of arbitrary slope on the reference-to-output map. The modification involves setting a reference slope in the algorithm, which, in extremum seeking, is implicitly set to zero.

Chapter 4

Discrete Time Extremum Seeking

This chapter treats a sinusoidal perturbation based extremum seeking scheme for discrete-time systems. The plant model and control algorithm have the same structure as in Chapter 1. Nevertheless, it turns out that the stability analysis of the discrete-time case is quite different from that of the continuous-time case because the frequency in discrete time systems lies between 0 and 2π and does not yield the time-scaling used for averaging in the continuous time case. By using two-time scale averaging theory [10] necessary in the discrete time case, we derive a sufficient condition under which the plant output exponentially converges to an $O(\alpha^2)$ neighborhood of the extremum value, where α is the magnitude of the modulation signal.

This chapter is organized as follows. Section 4.1 describes the discrete-time extremum seeking algorithm. Section 4.2 organizes the equations of the closed-loop system in a way convenient for stability analysis. Section 4.3 states and proves stability, and derives ultimate bounds on error signals, and Section 4.4 provides a simulation example and discussions.

4.1 Discrete-Time Extremum Seeking Control

The implementation is depicted in Figure 4.1 and takes the same structure as that of continuous time extremum seeking algorithm in Chapter 1. Both of the linear blocks, $F_i(z)$ and $F_o(z)$, are required to be exponentially stable. The high-pass filter $\frac{z-1}{z+h}$ is designed as $0 < h < 1$, and the modulation frequency ω is selected such that $\omega = a\pi$, $0 < |a| < 1$, and a is rational. Without loss of generality, the static nonlinear block $f(\theta)$ is assumed to have a minimum at $\theta = \theta^*$, and to be of the form

$$f(\theta) = f^* + (\theta - \theta^*)^2. \tag{4.1}$$

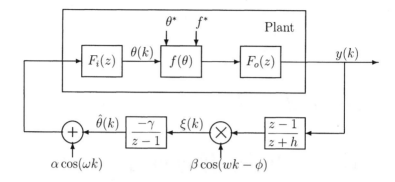

Figure 4.1: Extremum seeking control scheme for discrete-time systems

Cubic and higher order terms are omitted for notational convenience as they are negligible in local stability analysis via averaging.

4.2   Closed-Loop System

In the subsequent discussion, the following notation and definitions are used. A transfer function in front of a bracketed time function, such as $G(z)[u(k)]$, means a time-domain signal obtained as an output of $G(z)$ driven by $u(k)$. ε^{-k} denotes exponentially decaying terms.

The extremum seeking system depicted in Figure 4.1 is governed by the following equations:

$$y(k) \;=\; F_o(z)\Big[f^* + (\theta(k) - \theta^*)^2\Big], \tag{4.2}$$

$$\theta(k) \;=\; F_i(z)\Big[\alpha\cos(\omega k) - \frac{\gamma}{z-1}[\xi(k)]\Big], \tag{4.3}$$

$$\xi(k) \;=\; \beta\cos(\omega k - \phi)\frac{z-1}{z+h}[y(k)]. \tag{4.4}$$

For the convenience of analysis, the following terms are defined:

$$\theta_0(k) \;=\; F_i(z)[\alpha\cos(\omega k)], \tag{4.5}$$

$$\tilde{\theta}(k) \;=\; \theta^* - \theta(k) + \theta_0(k), \tag{4.6}$$

$$\tilde{y}(k) \;=\; y(k) - F_o(z)[f^*], \tag{4.7}$$

where $\tilde{\theta}(k)$ is the tracking error and $\tilde{y}(k)$ is the output error. Substitution of Eqns. (4.3) and (4.5) in Eqn. 4.6 yields

$$\tilde{\theta}(k) = \theta^* + \frac{\gamma}{z-1}F_i(z)[\xi(k)], \tag{4.8}$$

which can be transformed into a difference equation

$$\tilde{\theta}(k+1) = \tilde{\theta}(k) + \gamma F_i(z)[\xi(k)]. \tag{4.9}$$

Further, substitution for ξ from Eqn. (4.4) and for y from Eqn. (4.2) yields

$$\tilde{\theta}(k+1) - \tilde{\theta}(k) = \gamma F_i(z)\left[\beta c(\omega k)\frac{z-1}{z+h}F_o(z)[f^* + (\theta - \theta^*)^2]\right], \tag{4.10}$$

where

$$c(\omega k) \stackrel{\triangle}{=} \cos(\omega k - \phi).$$

Using $\theta - \theta^* = \theta_0 - \tilde{\theta}$ by rearrangement of Eqn. (4.6), we obtain

$$\begin{aligned}
\tilde{\theta}(k+1) - \tilde{\theta}(k) &= \gamma F_i(z)\left[\beta c(\omega k)\frac{z-1}{z+h}F_o(z)[f^* + (\theta_0 - \tilde{\theta})^2]\right] \\
&= \epsilon F_i(z)\left[c(\omega k)\frac{z-1}{z+h}F_o(z)[\tilde{\theta}^2 - 2\theta_0\tilde{\theta}]\right] \\
&+ \epsilon F_i(z)\left[c(\omega k)\frac{z-1}{z+h}F_o(z)[f^* + \theta_0^2]\right],
\end{aligned} \tag{4.11}$$

where

$$\epsilon \stackrel{\triangle}{=} \gamma\beta.$$

Applying the modulation Lemmas B.2, B.3, B.4 in succession to the term containing $2\theta_0\tilde{\theta}$ in Eqn. (4.11), we obtain

$$\begin{aligned}
&\epsilon F_i(z)\left[c(\omega k)\frac{z-1}{z+h}F_o(z)[-2\theta_0\tilde{\theta}]\right] \\
&= \alpha\epsilon F_i(z)\left[s(2\omega k)\mathbf{Im}\left\{F_i(e^{j\omega})\frac{e^{j\omega}z-1}{e^{j\omega}z+h}F_o(e^{j\omega}z)[\tilde{\theta}]\right\}\right. \\
&\qquad\left. -c(2\omega k)\mathbf{Re}\left\{F_i(e^{j\omega})\frac{e^{j\omega}z-1}{e^{j\omega}z+h}F_o(e^{j\omega}z)[\tilde{\theta}]\right\}\right] \\
&\quad -\alpha\epsilon F_i(z)\left[\mathbf{Re}\left\{e^{j\phi}F_i(e^{j\omega})\frac{e^{j\omega}z-1}{e^{j\omega}z+h}F_o(e^{j\omega}z)[\tilde{\theta}]\right\} + \varepsilon^{-k}\right], \tag{4.12}
\end{aligned}$$

where

$$\begin{aligned}
s(2\omega k) &\stackrel{\triangle}{=} \sin(2\omega k - \phi) \\
c(2\omega k) &\stackrel{\triangle}{=} \cos(2\omega k - \phi).
\end{aligned}$$

Finally, substituting Eqn. (4.12) in Eqn. (4.11) we obtain the whole closed-loop system

$$\tilde{\theta}(k+1) - \tilde{\theta}(k) = \epsilon\Big(L(z)[\tilde{\theta}] + \Phi_1(k) + \Phi_2(k)\Big) + \delta(k), \tag{4.13}$$

where

$$
L(z) = -\frac{\alpha}{2}F_i(z)\Big(e^{j\phi}M(z, e^{j\omega}) + e^{-j\phi}M(z, e^{-j\omega})\Big),
$$

$$
\Phi_1(k) = \alpha F_i(z)\Big[s(2\omega k)\mathbf{Im}\big\{M(z, e^{j\omega})[\tilde{\theta}]\big\} - c(2\omega k)\mathbf{Re}\big\{M(z, e^{j\omega})[\tilde{\theta}]\big\}\Big],
$$

$$
\Phi_2(k) = F_i(z)\Big[c(\omega k)\frac{z-1}{z+h}F_o(z)[\tilde{\theta}^2]\Big],
$$

$$
\delta(k) = \epsilon F_i(z)\Big[c(\omega k)\frac{z-1}{z+h}F_o(z)[f^* + \theta_0^2] + \alpha\varepsilon^{-k}\Big],
$$

$$
M(z, e^{j\omega}) = F_i(e^{j\omega})\frac{e^{j\omega}z-1}{e^{j\omega}z+h}F_o(e^{j\omega}z).
$$

The various terms in Eqn. (4.13) can be characterized in view of $\tilde{\theta}$ as follows: $L(z)[\tilde{\theta}]$ is linear time-invariant; $\Phi_1(k)$ is linear time-varying; $\Phi_2(k)$ is nonlinear time-varying; and $\delta(k)$ is time-varying, independent of $\tilde{\theta}$, and found to satisfy the following property:

Lemma 4.1 $\delta(k)$ *exponentially converges to an $O(\epsilon\alpha^2)$ neighborhood of zero:*

$$
|\delta(k)| \le \varepsilon^{-k} + \kappa_1\epsilon\alpha^2, \tag{4.14}
$$

where κ_1 is a constant.

Proof. The term $\theta_0^2(k)$ in $\delta(k)$ is calculated as

$$
\theta_0^2(k) = \frac{1}{2}\alpha^2\big|F_i(e^{j\omega})\big|^2\Big(1 + \cos(2k\omega + 2\psi_1)\Big) + \varepsilon^{-k}, \tag{4.15}
$$

where $\psi_1 = \angle\big(F_i(e^{j\omega})\big)$. Then, $\delta(k)$ is rearranged as

$$
\delta(k) = \delta_1(k) + \delta_2(k), \tag{4.16}
$$

where

$$
\delta_1(k) = \epsilon F_i(z)\Big[\cos(\omega k - \phi)\frac{z-1}{z+h}F_o(z)\big[f^* + \frac{1}{2}\alpha^2|F_i(e^{j\omega})|^2\big]\Big] + \varepsilon^{-k}, \tag{4.17}
$$

$$
\delta_2(k) = \frac{1}{2}\epsilon\alpha^2\big|F_i(e^{j\omega})\big|^2
$$
$$
\times F_i(z)\Big[\cos(\omega k - \phi)\frac{z-1}{z+h}F_o(z)\big[\cos(2k\omega + 2\psi_1)\big]\Big]. \tag{4.18}
$$

Since the high-pass filter $\frac{z-1}{z+h}$ has zero DC gain, $\delta_1(k)$ in Eqn. (4.16) contains only exponentially decaying terms. On the other hand, by using Lemma B.1, $\delta_2(k)$ is calculated as

$$
\delta_2(k) = \frac{1}{2}\epsilon\alpha^2 c_1 F_i(z)\Big[\cos(\omega k - \phi)\cos(2\omega k + \psi_2)\Big] \tag{4.19}
$$

$$= \frac{1}{4}\epsilon\alpha^2 c_1 F_i(z)\Big[\cos(3\omega k - \phi + \psi_2) + \cos(\omega k + \phi + \psi_2)\Big] \qquad (4.20)$$

$$= \frac{1}{4}\epsilon\alpha^2 c_1 \Big(\big|F_i(e^{j3\omega})\big|\cos(3\omega k - \phi + \psi_3)$$

$$+ \big|F_i(e^{j\omega})\big|\cos(\omega k + \phi + \psi_4)\Big) \qquad (4.21)$$

$$\leq \kappa_1 \epsilon\alpha^2, \qquad (4.22)$$

where $c_1 = \big|F_i(e^{j\omega})\big|^2 \big|\frac{e^{j2\omega}-1}{e^{j2\omega}+h}\big| \big|F_o(e^{j2\omega})\big|$, $\psi_2 = 2\psi_1 + \angle\big(F_o(e^{j2\omega})\big) + \angle\big(\frac{e^{j2\omega}-1}{e^{j2\omega}+h}\big)$, $\psi_3 = \psi_2 + \angle\big(F_i(e^{j3\omega})\big)$, $\psi_4 = \psi_2 + \angle\big(F_i(e^{j\omega})\big)$ and $\kappa_1 = \frac{1}{4}c_1\big(\big|F_i(e^{j3\omega})\big| + \big|F_i(e^{j\omega})\big|\big)$.

$$\text{Q.E.D.}$$

From Lemma 4.1, it is clear that the bound on $\delta(k)$ can be adjusted by the magnitude of the modulation signal α independently of ϵ. By exploiting this property of $\delta(k)$, we present a stability analysis for the system in Eqn. 4.13 with two steps. At first, regarding $\delta(k)$ as a perturbation, we analyze the system in Eqn. (4.13) without $\delta(k)$. Then, we consider the whole system including $\delta(k)$.

4.3   Stability Analysis

First, we consider the homogeneous part of the $\tilde{\theta}$-error system Eqn. (4.13)

$$\tilde{\theta}(k+1) - \tilde{\theta}(k) = \epsilon\Big(L(z)[\tilde{\theta}] + \Phi_1(k) + \Phi_2(k)\Big), \qquad (4.23)$$

which depends on time k periodically. The following theorem presents a sufficient condition under which the $\tilde{\theta}$-error system Eqn. (4.23) is locally exponentially stable at the origin:

Theorem 4.2 *If* $F_i(1)\mathbf{Re}\big\{e^{j\phi}F_i(e^{j\omega})\frac{e^{j\omega}-1}{e^{j\omega}+h}F_o(e^{j\omega})\big\} > 0$, *then there exists a positive constant* ϵ^* *such that the state-space realization of the* $\tilde{\theta}$-error system *Eqn. (4.23) is locally exponentially stable at the origin for all* $0 < \epsilon(= \gamma\beta) \leq \epsilon^*$.

Proof: Since $\Phi_1(k)$ and $\Phi_2(k)$ in Eqn. (4.23) take the same structure as $G(z)\big[\cos(\omega k - \phi)H(z)[v(k)]\big]$ in Lemma B.5, we can choose minimal state space realizations of $L(z)$, $\Phi_1(k)$, and $\Phi_2(k)$ as (A_1, B_1, C_1, D_1), $(A_2(k), B_2(k), C_2(k), D_2(k))$, and $(A_3(k), B_3(k), C_3(k), D_3(k))$, respectively. Moreover, since all of the poles in $L(z)$, $\Phi_1(k)$, and $\Phi_2(k)$ are inside the unit circle, A_1, $A_2(k)$, and $A_3(k)$ are exponentially stable. Now, the $\tilde{\theta}$-error system Eqn. (4.23) can be transformed into a state space form

$$\begin{aligned} x'(k+1) &= A(k)x'(k) + h(k, \tilde{\theta}(k)) \qquad &(4.24) \\ \tilde{\theta}(k+1) &= \tilde{\theta}(k) + \epsilon f'(k, \tilde{\theta}(k), x'(k)), \qquad &(4.25) \end{aligned}$$

where

$$A(k) = \begin{bmatrix} A_1 & 0 & 0 \\ 0 & A_2(k) & 0 \\ 0 & 0 & A_3(k) \end{bmatrix}$$

$$h(k, \tilde{\theta}(k)) = \left[B_1^T \tilde{\theta} \mid B_2^T(k)\tilde{\theta} \mid B_3^T(k)\tilde{\theta}^2 \right]^T$$

$$f'(k, \tilde{\theta}(k), x'(k)) = D_1\tilde{\theta} + D_2\tilde{\theta} + D_3\tilde{\theta}^2 + \left[C_1 \mid C_2(k) \mid C_3(k) \right] x'(k).$$

Since $A(k)$ is exponentially stable, the state space form Eqns. (4.24) and (4.25) is adequate for the application of the two-time scale averaging theory [10]. Define the function

$$w(k, \tilde{\theta}) = \sum_{i=0}^{k-1} \Psi(k, i+1)h(i, \tilde{\theta}), \tag{4.26}$$

where $\Psi(k, i) = \prod\limits_{l=i}^{k-1} A(i + k - 1 - l)$, and construct the transformation

$$x(k) = x'(k) - w(k, \tilde{\theta}). \tag{4.27}$$

Then, the transformed system is represented as

$$x(k+1) = A(k)x(k) + \epsilon g(k, \tilde{\theta}, x) \tag{4.28}$$
$$\tilde{\theta}(k+1) = \tilde{\theta}(k) + \epsilon f(k, \tilde{\theta}, x), \tag{4.29}$$

where

$$g(k, \tilde{\theta}, x) = -\left(\int_0^1 \frac{\partial w}{\partial \tilde{\theta}}(k+1, s\tilde{\theta}(k+1) + (1-s)\tilde{\theta}(k))ds \right) f'(k, \tilde{\theta}, x + w(k, \tilde{\theta}))$$

$$f(k, \tilde{\theta}, x) = f'(k, \tilde{\theta}(k), x + w(k, \tilde{\theta})).$$

The averaged system of Eqn. (4.29) is defined by

$$\tilde{\theta}_{av}(k+1) = \tilde{\theta}_{av}(k) + \epsilon f_{av}(\tilde{\theta}_{av}(k)), \tag{4.30}$$

where f_{av} is calculated by the averaging operator $\mathbf{AVG}\{\cdot\}$ [10] defined as

$$f_{av}(\tilde{\theta}) \stackrel{\triangle}{=} \mathbf{AVG}\left\{ f(k, \tilde{\theta}, 0) \right\}$$

$$\stackrel{\triangle}{=} \lim_{T \to \infty} \frac{1}{T} \sum_{k=s+1}^{s+T} f(k, \tilde{\theta}, 0).$$

On the other hand, $f(k, \tilde{\theta}, 0)$ can be reconverted into $\mathcal{Z}$-domain as follows:

$$f(k, \tilde{\theta}, 0) = f'(k, \tilde{\theta}, w(k, \tilde{\theta}))$$
$$= D_1\tilde{\theta} + D_2\tilde{\theta} + D_3\tilde{\theta}^2$$
$$+ \left[C_1 \mid C_2(k) \mid C_3(k) \right] \sum_{i=0}^{k-1} \Psi(k, i+1)\left[B_1^T \tilde{\theta} \mid B_2^T(i)\tilde{\theta} \mid B_3^T(i)\tilde{\theta}^2 \right]^T$$
$$= L(z)[\tilde{\theta}] + \Phi_1(k) + \Phi_2(k),$$

where $\tilde{\theta}$ is regarded as a constant. Hence $f_{av}(\tilde{\theta})$ can be reformulated as

$$f_{av}(\tilde{\theta}) = \mathbf{AVG}\big\{ L(z)[\tilde{\theta}] + \Phi_1(k) + \Phi_2(k) \big\}. \tag{4.31}$$

Using Lemma B.1 and regarding $\tilde{\theta}$ as a constant leads to the following derivations:

$$\mathbf{AVG}\big\{\Phi_1(k)\big\} = \mathbf{AVG}\big\{ aF_i(z)\big[s(2\omega k)\mathbf{Im}\{M(z,e^{j\omega})[u_s(k)]\}$$
$$-c(2\omega k)\mathbf{Re}\{M(z,e^{j\omega})[u_s(k)]\}\big]\tilde{\theta}\big\} = 0, \tag{4.32}$$

$$\mathbf{AVG}\big\{\Phi_2(k)\big\} = \mathbf{AVG}\big\{ F_i(z)\big[c(\omega k)\frac{z-1}{z+h}F_o(z)[u_s(k)]\big]\tilde{\theta}^2\big\} = 0, \tag{4.33}$$

where $u_s(k)$ denotes the unit step sequence, $s(2\omega k) = \sin(2\omega k - \phi)$, $c(2\omega k) = \cos(2\omega k - \phi)$, and $c(\omega k) = \cos(\omega k - \phi)$. Hence,

$$f_{av}(\tilde{\theta}) = \mathbf{AVG}\big\{ L(z)[\tilde{\theta}]\big\}$$
$$= \mathbf{AVG}\big\{ -\frac{\alpha}{2}F_i(z)\big(e^{j\phi}M(z,e^{j\omega}) + e^{-j\phi}M(z,e^{-j\omega})\big)[u_s(k)]\tilde{\theta}\big\}$$
$$= -\kappa_2\alpha\tilde{\theta}, \tag{4.34}$$

where

$$\kappa_2 = \frac{1}{2}F_i(1)\mathbf{Re}\big\{ e^{j\phi}F_i(e^{j\omega})\frac{e^{j\omega}-1}{e^{j\omega}+h}F_o(e^{j\omega})\big\}$$
$$= \frac{1}{2}F_i(1)\big| F_i(e^{j\omega})\frac{e^{j\omega}-1}{e^{j\omega}+h}F_o(e^{j\omega})\big| \cos(\psi_M + \phi),$$

and $\psi_M = \angle\big(F_i(e^{j\omega})\frac{e^{j\omega}-1}{e^{j\omega}+h}F_o(e^{j\omega})\big)$. Substituting Eqn. (4.34) into Eqn. (4.30) results in the averaged system

$$\tilde{\theta}_{av}(k+1) = (1 - \kappa_2\epsilon\alpha)\tilde{\theta}_{av}(k), \tag{4.35}$$

where if $\kappa_2 > 0$, $\tilde{\theta}_{av}$ is exponentially stable for all $0 < \epsilon < \frac{2}{\kappa_2\alpha}$. Consequently, according to Theorem 2.2.4 in [10], this theorem is proved. Q.E.D.

It is observed from the sufficient condition of Theorem 4.2 that the local exponential stability of Eqn. (4.23) is closely related to positive realness of linear parts of the plant but only at the modulation frequency ω. This is a very mild condition.

Now, we consider the stability of the overall system Eqn. (4.13). For this purpose, it is necessary to investigate the perturbed averaged system

$$\tilde{\theta}_{av}(k+1) = (1 - \kappa_2\epsilon\alpha)\tilde{\theta}_{av}(k) + \delta(k). \tag{4.36}$$

Since $|\delta(k)| \leq \varepsilon^{-k} + \kappa_1 \epsilon \alpha^2$ from Lemma 4.1, it is obvious that $\tilde{\theta}_{av}(k)$ in Eqn. (4.36) exponentially converges to an $O(\alpha)$ neighborhood of zero. On the other hand, it is known from [10] and [98] that the exponential convergence rate of $\tilde{\theta}$ in the original system Eqn. (4.13) tends to that of $\tilde{\theta}_{av}$ in the averaged system, as ϵ tends to zero. Therefore, we can conclude the following theorem.

Theorem 4.3 *Suppose that the conditions of Theorem 4.2 are satisfied. Then, for sufficiently small α, there exists ϵ_1^*, $0 < \epsilon_1^* \leq \epsilon^*$, such that $\tilde{\theta}$ in the original system Eqn. (4.13) locally exponentially converges to an $O(\alpha)$ neighborhood of zero for all $0 < \epsilon \leq \epsilon_1^*$.*

Note that the requirement that ϵ be small translates into a requirement that γ be small. With the result of Theorem 4.3, the convergence property of the output error $\tilde{y}(k)$ is described as:

Corollary 4.4 *Under the conditions of Theorem 4.3, the output error $\tilde{y}(k)$ defined in Eqn. (4.7) locally exponentially converges to an $O(\alpha^2)$ neighborhood of zero.*

Proof: We have that

$$\tilde{y}(k) = F_o(z)\left[(\theta - \theta^*)^2\right] = F_o(z)\left[(\tilde{\theta} - \theta_0)^2\right], \tag{4.37}$$

where $\tilde{\theta}$ locally exponentially converges to an $O(\alpha)$ neighborhood of zero from Theorem 4.3 and θ_0 exponentially converges to an $O(\alpha)$ neighborhood of zero. Hence, $\tilde{y}(k)$ locally exponentially converges to an $O(\alpha^2)$ neighborhood of zero.   Q.E.D.

4.4   Example

In order to test the feasibility of the proposed extremum seeking algorithm, we conduct a simulation study for a plant with transfer functions

$$F_i(z) = \frac{z + 0.4}{(z + 0.5)(z + 0.6)} \text{ and } F_o(z) = \frac{z - 0.2}{z + 0.6}. \tag{4.38}$$

Other design parameters are selected as: $\theta^* = 3$, $f^* = 2$, $h = 0.9$, $\alpha = 0.05$, $\beta = 0.05$, and $\phi = 0$. Simulation is conducted for $\omega = \frac{\pi}{1.1}$ and $\omega = \frac{\pi}{1.5}$, so that it can be calculated that $\left|M(e^{j\frac{\pi}{1.1}})\right| = 4.57$, $\angle\left(M(e^{j\frac{\pi}{1.1}})\right) = -0.75$ rad, $\left|M(e^{j\frac{\pi}{1.5}})\right| = 2.68$, $\angle\left(M(e^{j\frac{\pi}{1.5}})\right) = 0.93$ rad, and $F_i(1) = 0.58$, where $M(e^{j\omega}) = F_i(e^{j\omega})\frac{e^{j\omega}-1}{e^{j\omega}+h}F_o(e^{j\omega})$. Since $\cos\left(\angle\left(M(e^{j\frac{\pi}{1.1}})\right)\right) > 0$, $\cos\left(\angle\left(M(e^{j\frac{\pi}{1.5}})\right)\right) > 0$, and $F_i(1) > 0$, the sufficient condition of Theorem 4.2 is satisfied for both $\omega =$

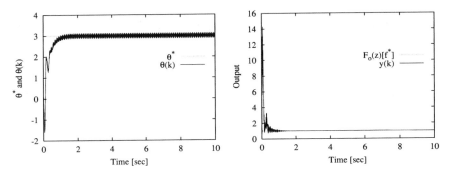

Figure 4.2: Responses for $\omega = \frac{\pi}{1.1}$ rad/sample and $\gamma = 0.6$

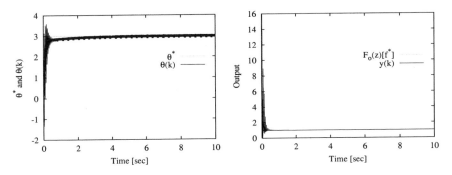

Figure 4.3: Responses for $\omega = \frac{\pi}{1.5}$ rad/sample and $\gamma = 2.1$

$\frac{\pi}{1.1}$ and $\frac{\pi}{1.5}$. Accordingly, it is certain that the system is exponentially stable, which is illustrated in Figures 4.2 and 4.3. It is also shown from Figures 4.2 and 4.3 that $\theta(k)$ converges to θ^* with larger magnitude of oscillation than that in the convergence of $y(k)$ to $F_o(z)[f^*]$. This observation illustrates the results of Theorem 4.3 and Corollary 4.4 that $\tilde{\theta}(k)$ and $\tilde{y}(k)$ locally exponentially converge to $O(\alpha)$ and $O(\alpha^2)$ neighborhoods of zero, respectively.

Notes and References

This chapter is based upon Choi et al. [30]. By using the two-time scale averaging theory [10], the authors derived a very mild sufficient condition under which the system output converges exponentially to an $O(\alpha^2)$ neighborhood of the extremum value. The sufficient condition is related to positive realness of linear parts of the plant but only at the modulation frequency ω. Future study subjects should include: development of a method to improve and analyze the transient performance; rejection of measurement noise; tracking of time-varying f^* and θ^*; and practical design guidelines for selecting modulation signal frequency ω, phase shift of demodulation signal ϕ, and various other gains. Other works on discrete time extremum seeking have essentially been based upon traditional search [122] and nonlinear optimization methods [107] which require assuming a fast plant, or, conversely, slow convergence to the optimum.

Chapter 5

Nonlinear Analysis

In this chapter we present the general problem where the nonlinearity with an extremum arises as a reference-to-output *equilibrium* map of a general nonlinear (non-affine in control) system. Such a system is assumed to be stable or stabilizable at each of these equilibria by a local feedback controller. We consider the single parameter problem in this chapter. In relation to Chapter 1, this chapter can be viewed as a generalization where $F_o(s)$ is allowed to be nonlinear (while $F_i(s)$ is required to be constant, or very fast).

An example of such a problem is the compressor stall and surge problem, presented in Chapters 11 and 12. In fact, the equilibrium map with the extremum there is nonunique–it has multiple bifurcating branches.

We employ the tools of averaging and singular perturbations to show that solutions of the closed-loop system converge to a small neighborhood of the extremum of the equilibrium map (in this and the following chapter, the problem is posed to seek maxima). The size of the neighborhood is inversely proportional to the adaptation gain and the amplitude and frequency of a periodic signal used to achieve extremum seeking. Our analysis highlights a fundamentally nonlinear mechanism of stabilization in an extremum seeking loop. After stating the problem in Section 5.1 and giving the extremum seeking scheme in Section 5.2, our proof is presented in Sections 5.3 and 5.4.

5.1 Extremum Seeking: Problem Statement

Consider the nonlinear model

$$\dot{x} = f(x, u) \tag{5.1}$$
$$y = h(x), \tag{5.2}$$

where $x \in \mathbb{R}^n$ is the state, $u \in \mathbb{R}$ is the input, $y \in \mathbb{R}$ is the output, and $f : \mathbb{R}^n \times \mathbb{R} \to \mathbb{R}^n$ and $h : \mathbb{R}^n \to \mathbb{R}$ are smooth. Suppose that we know a smooth control law

$$u = \alpha(x, \theta) \tag{5.3}$$

parameterized by a scalar parameter θ. The closed-loop system

$$\dot{x} = f\left(x, \alpha(x, \theta)\right) \tag{5.4}$$

then has equilibria parameterized by θ. We make the following assumptions about the closed-loop system.

Assumption 5.1 *There exists a smooth function* $l : \mathbb{R} \to \mathbb{R}^n$ *such that*

$$f\left(x, \alpha(x, \theta)\right) = 0 \qquad \text{if and only if} \qquad x = l(\theta). \tag{5.5}$$

Assumption 5.2 *For each* $\theta \in \mathbb{R}$, *the equilibrium* $x = l(\theta)$ *of the system (5.4) is locally exponentially stable.*

Hence, we assume that we have a control law (5.3) which *exponentially stabilizes any of the equilibria that θ may produce*. Except for the requirement that Assumption 5.2 holds *for any $\theta \in \mathbb{R}$* (which we impose only for notational convenience and can easily relax to an *interval* in $\mathbb{R}$), this assumption is not restrictive. It simply means that we have a control law designed for local stabilization and this control law need not be based on modeling knowledge of either $f(x, u)$ or $l(\theta)$.

The next assumption is central to the problem of peak seeking.

Assumption 5.3 *There exists* $\theta^* \in \mathbb{R}$ *such that*

$$(h \circ l)'(\theta^*) = 0 \tag{5.6}$$
$$(h \circ l)''(\theta^*) < 0. \tag{5.7}$$

Thus, we assume that the output equilibrium map $y = h\left(l(\theta)\right)$ has a *maximum* at $\theta = \theta^*$. Our objective is to develop a feedback mechanism which maximizes the steady state value of y but without requiring knowledge of either θ^* or the functions h and l. Our assumption that $h \circ l$ has a maximum is without loss of generality—the case with a minimum would be treated identically by replacing y by $-y$ in the subsequent feedback design.

5.2   Extremum Seeking Scheme

The extremum seeking scheme is shown in Figure 5.1. The low pass filter $\frac{\omega_l}{s+\omega_l}$, which did not appear in previous chapters, is not necessary, but it is helpful in filtering out a $\cos 2\omega t$ signal after the multiplier (demodulator). The design parameters are selected as

$$\omega_h = \omega\omega_H = \omega\delta\omega'_H = O(\omega\delta) \tag{5.8}$$
$$\omega_l = \omega\omega_L = \omega\delta\omega'_L = O(\omega\delta) \tag{5.9}$$
$$k = \omega K = \omega\delta K' = O(\omega\delta), \tag{5.10}$$

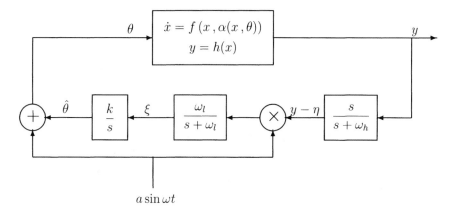

Figure 5.1: Extremum seeking scheme.

where ω and δ are small positive constants and ω'_H, ω'_L, and K' are $O(1)$ positive constants. As it will become apparent later, a also needs to be small.

From (5.8) and (5.9) we see that the cut-off frequencies of the filters need to be lower than the frequency of the perturbation signal. In addition, the adaptation gain k needs to be small. Thus, the overall feedback system has three time scales:

- fastest—the plant with the stabilizing controller,

- medium—the periodic perturbation,

- slow—the filters in the extremum seeking scheme.

The analysis that follows treats first the static case from Figure 5.2 using the method of *averaging* (Section 5.3). Then we use the *singular perturbation* method (Section 5.4) for the full system in Figure 5.1.

Before we start our analysis, we summarize the system in Figure 5.1 as

$$\dot{x} = f\left(x, \alpha(x, \hat{\theta} + a\sin\omega t)\right) \tag{5.11}$$

$$\dot{\hat{\theta}} = k\xi \tag{5.12}$$

$$\dot{\xi} = -\omega_l\xi + \omega_l(y - \eta)a\sin\omega t \tag{5.13}$$

$$\dot{\eta} = -\omega_h\eta + \omega_h y. \tag{5.14}$$

Let us introduce the new coordinates

$$\tilde{\theta} = \hat{\theta} - \theta^* \tag{5.15}$$

$$\tilde{\eta} = \eta - h \circ l(\theta^*). \tag{5.16}$$

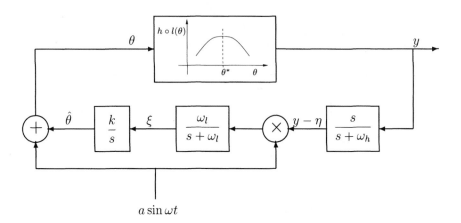

Figure 5.2: If perturbation $a \sin \omega t$ is slow, the plant can be viewed as a static map.

Then, in the time scale $\tau = \omega t$, the system (5.11)–(5.14) is rewritten as

$$\omega \frac{dx}{d\tau} = f\left(x, \alpha(x, \theta^* + \tilde{\theta} + a \sin \tau)\right) \tag{5.17}$$

$$\frac{d}{d\tau} \begin{bmatrix} \tilde{\theta} \\ \xi \\ \tilde{\eta} \end{bmatrix} = \delta \begin{bmatrix} K'\xi \\ -\omega'_L \xi + \omega'_L \left(h(x) - h \circ l(\theta^*) - \tilde{\eta}\right) a \sin \tau \\ -\omega'_H \tilde{\eta} + \omega'_H \left(h(x) - h \circ l(\theta^*)\right) \end{bmatrix} . \tag{5.18}$$

5.3 Averaging Analysis

The first step in our analysis is to study the system in Figure 5.2. We "freeze" x in (5.17) at its "equilibrium" value

$$x = l\left(\theta^* + \tilde{\theta} + a \sin \tau\right) \tag{5.19}$$

and substitute it into (5.18), getting the "reduced system"

$$\frac{d}{d\tau} \begin{bmatrix} \tilde{\theta}_r \\ \xi_r \\ \tilde{\eta}_r \end{bmatrix} = \delta \begin{bmatrix} K'\xi_r \\ -\omega'_L \xi_r + \omega'_L \left(\nu \left(\tilde{\theta}_r + a \sin \tau\right) - \tilde{\eta}_r\right) a \sin \tau \\ -\omega'_H \tilde{\eta}_r + \omega'_H \nu \left(\tilde{\theta}_r + a \sin \tau\right) \end{bmatrix} , \tag{5.20}$$

where

$$\nu\left(\tilde{\theta}_r + a \sin \tau\right) = h \circ l\left(\theta^* + \tilde{\theta}_r + a \sin \tau\right) - h \circ l(\theta^*) . \tag{5.21}$$

In view of Assumption 5.3, it is obvious that

$$\nu(0) = 0 \tag{5.22}$$

$$\nu'(0) = (h \circ l)'(\theta^*) = 0 \tag{5.23}$$
$$\nu''(0) = (h \circ l)''(\theta^*) < 0. \tag{5.24}$$

The system (5.20) is in the form to which the averaging method is applicable. The average model of (5.20) is

$$\frac{d}{d\tau} \begin{bmatrix} \tilde{\theta}_r^a \\ \xi_r^a \\ \tilde{\eta}_r^a \end{bmatrix} = \delta \begin{bmatrix} K'\xi_r^a \\ -\omega_L'\xi_r^a + \frac{\omega_L'}{2\pi}a \int_0^{2\pi} \nu \left(\tilde{\theta}_r^a + a\sin\sigma\right)\sin\sigma d\sigma \\ -\omega_H'\tilde{\eta}_r^a + \frac{\omega_H'}{2\pi}\int_0^{2\pi} \nu \left(\tilde{\theta}_r^a + a\sin\sigma\right)d\sigma \end{bmatrix}. \tag{5.25}$$

First we need to determine the average equilibrium $\left(\tilde{\theta}_r^{a,e}, \xi_r^{a,e}, \tilde{\eta}_r^{a,e}\right)$ which satisfies

$$\xi_r^{a,e} = 0 \tag{5.26}$$
$$\int_0^{2\pi} \nu \left(\tilde{\theta}_r^{a,e} + a\sin\sigma\right)\sin\sigma d\sigma = 0 \tag{5.27}$$
$$\tilde{\eta}_r^{a,e} = \frac{1}{2\pi}\int_0^{2\pi} \nu \left(\tilde{\theta}_r^{a,e} + a\sin\sigma\right)d\sigma. \tag{5.28}$$

By postulating $\tilde{\theta}_r^{a,e}$ in the form

$$\tilde{\theta}_r^{a,e} = b_1 a + b_2 a^2 + O(a^3), \tag{5.29}$$

substituting in (5.27), using (5.22) and (5.23), integrating, and equating the like powers of a, we get $\nu''(0)b_1 = 0$ and $\nu''(0)b_2 + \frac{1}{8}\nu'''(0) = 0$, which implies that

$$\tilde{\theta}_r^{a,e} = -\frac{\nu'''(0)}{8\nu''(0)}a^2 + O(a^3). \tag{5.30}$$

Another round of lengthy calculations applied to (5.28) yields

$$\tilde{\eta}_r^{a,e} = \frac{\nu''(0)}{4}a^2 + O(a^3). \tag{5.31}$$

Thus, the equilibrium of the average model (5.25) is

$$\begin{bmatrix} \tilde{\theta}_r^{a,e} \\ \xi_r^{a,e} \\ \tilde{\eta}_r^{a,e} \end{bmatrix} = \begin{bmatrix} -\frac{\nu'''(0)}{8\nu''(0)}a^2 + O(a^3) \\ 0 \\ \frac{\nu''(0)}{4}a^2 + O(a^3) \end{bmatrix}. \tag{5.32}$$

The Jacobian of (5.25) at $\left(\tilde{\theta}, \xi, \tilde{\eta}\right)_r^{a,e}$ is

$$J_r^a = \delta \begin{bmatrix} 0 & K' & 0 \\ \frac{\omega_L'}{2\pi}a \int_0^{2\pi} \nu' \left(\tilde{\theta}_r^{a,e} + a\sin\sigma\right)\sin\sigma d\sigma & -\omega_L' & 0 \\ \frac{\omega_H'}{2\pi}\int_0^{2\pi} \nu' \left(\tilde{\theta}_r^{a,e} + a\sin\sigma\right)d\sigma & 0 & -\omega_H' \end{bmatrix}. \tag{5.33}$$

Since J_r^a is block-lower-triangular we easily see that it will be Hurwitz if and only if

$$\int_0^{2\pi} \nu' \left(\tilde{\theta}_r^{a,e} + a \sin \sigma \right) \sin \sigma d\sigma < 0 \,. \tag{5.34}$$

More calculations that use (5.22) and (5.23) give

$$\int_0^{2\pi} \nu' \left(\tilde{\theta}_r^{a,e} + a \sin \sigma \right) \sin \sigma d\sigma = \pi \nu''(0) a + O(a^2) \,. \tag{5.35}$$

By substituting (5.35) into (5.33) we get

$$det \left(\lambda \mathrm{I} - J_r^a \right) = \left(\lambda^2 + \delta \omega_L' \lambda - \frac{\delta^2 \omega_L' K'}{2} \nu''(0) a^2 + O(\delta^2 a^3) \right) \left(\lambda + \delta \omega_H' \right) \,, \tag{5.36}$$

which, in view of (5.24), proves that J_r^a is Hurwitz for sufficiently small a. This, in turn, implies that the equilibrium (5.32) of the average system (5.25) is exponentially stable for a sufficiently small a. Then, according to the Averaging Theorem [64, Theorem 8.3] we have the following result.

Theorem 5.4 *Consider the system (5.20) under Assumption 5.3. There exist $\bar{\delta}$ and $\bar{a}$ such that for all $\delta \in (0, \bar{\delta})$ and $a \in (0, \bar{a})$ the system (5.20) has a unique exponentially stable periodic solution $\left(\tilde{\theta}_r^{2\pi}(\tau), \xi_r^{2\pi}(\tau), \tilde{\eta}_r^{2\pi}(\tau) \right)$ of period 2π and this solution satisfies*

$$\left\| \left[\begin{array}{c} \tilde{\theta}_r^{2\pi}(\tau) + \frac{\nu'''(0)}{8\nu''(0)} a^2 \\ \xi_r^{2\pi}(\tau) \\ \tilde{\eta}_r^{2\pi}(\tau) - \frac{\nu''(0)}{4} a^2 \end{array} \right] \right\| \le O(\delta) + O(a^3) \,, \qquad \forall \tau \ge 0 \,. \tag{5.37}$$

This result implies that all solutions $\left(\tilde{\theta}_r(\tau), \xi_r(\tau), \tilde{\eta}_r(\tau) \right)$, and, in particular, their $\tilde{\theta}_r(\tau)$-components, converge to an $O(\delta + a^2)$-neighborhood of the origin. It is important to interpret this result in terms of the system in Figure 5.2. Since $y = h \circ l \left(\theta^* + \tilde{\theta}_r(\tau) + a \sin \tau \right)$ and $(h \circ l)' (\theta^*) = 0$, we have

$$y - h \circ l(\theta^*) = (h \circ l)'' (\theta^*) \left(\tilde{\theta}_r + a \sin \tau \right)^2 + O \left(\left(\tilde{\theta}_r + a \sin \tau \right)^3 \right) \,, \tag{5.38}$$

where

$$\tilde{\theta}_r + a \sin \tau = \left(\tilde{\theta}_r - \tilde{\theta}_r^{2\pi} \right) + \left(\tilde{\theta}_r^{2\pi} + \frac{(h \circ l)''' (\theta^*)}{8 (h \circ l)'' (\theta^*)} a^2 \right)$$
$$- \frac{(h \circ l)''' (\theta^*)}{8 (h \circ l)'' (\theta^*)} a^2 + a \sin \tau \,. \tag{5.39}$$

Since the first term converges to zero, the second term is $O \left(\delta + a^3 \right)$, the third term is $O \left(a^2 \right)$ and the fourth term is $O(a)$, then

$$\limsup_{\tau \to \infty} \left| \tilde{\theta}_r(\tau) + a \sin \tau \right| = O(a + \delta) \,. \tag{5.40}$$

Thus, (5.38) yields

$$\limsup_{\tau \to \infty} |y(\tau) - h \circ l(\theta^*)| = O\left(a^2 + \delta^2\right) . \tag{5.41}$$

The last expression characterizes the asymptotic performance of the extremum seeking scheme in Figure 5.2 and explains why it is not only important that the periodic perturbation be small but also that the cut-off frequencies of the filters and the adaptation gain k be low.

Another important conclusion can be drawn from (5.37). The solution $\tilde{\theta}_r(\tau)$ will converge $O\left(\delta + a^3\right)$-close to $-\dfrac{(h \circ l)''' (\theta^*)}{8\,(h \circ l)'' (\theta^*)} a^2$. Since $(h \circ l)'' (\theta^*) < 0$, the sign of this quantity depends on the sign of $(h \circ l)''' (\theta^*)$. If $(h \circ l)''' (\theta^*) > 0$ (respectively, < 0), then the curve $h \circ l(\theta)$ will be more "flat" on the right (respectively, left) side of $\theta = \theta^*$. Since $\tilde{\theta}_r$ will have an offset in the direction of $\operatorname{sgn}\left\{(h \circ l)''' (\theta^*)\right\}$, then $\tilde{\theta}_r(t)$ will converge to the "flatter" side of $h \circ l(\theta)$. This is precisely what we want—to be on the side where $h \circ l(\theta)$ is less sensitive to variations in θ and closer to its maximum value.

5.4 Singular Perturbation Analysis

Now we address the full system in Figure 5.1 whose state space model is given by (5.17) and (5.18) in the time scale $\tau = \omega t$. To make the notation in our further analysis compact, we write (5.18) as

$$\frac{dz}{d\tau} = \delta G\left(\tau, x, z\right) , \tag{5.42}$$

where $z = \left(\tilde{\theta}, \xi, \tilde{\eta}\right)$. By Theorem 5.4, there exists an exponentially stable periodic solution $z_r^{2\pi}(\tau)$ such that

$$\frac{dz_r^{2\pi}(\tau)}{d\tau} = \delta G\left(\tau, L\left(\tau, z_r^{2\pi}(\tau)\right), z_r^{2\pi}(\tau)\right) , \tag{5.43}$$

where $L(\tau, z) = l\left(\theta^* + \tilde{\theta} + a \sin \tau\right)$. To bring the system (5.17) and (5.42) into the *standard singular perturbation form*, we shift the state z using the transformation

$$\tilde{z} = z - z_r^{2\pi}(\tau) \tag{5.44}$$

and get

$$\frac{d\tilde{z}}{d\tau} = \delta \tilde{G}\left(\tau, x, \tilde{z}\right) \tag{5.45}$$

$$\omega \frac{dx}{d\tau} = \tilde{F}\left(\tau, x, \tilde{z}\right) , \tag{5.46}$$

where

$$\tilde{G}\left(\tau,x,\tilde{z}\right) \;\; = \;\; G\left(\tau,x,\tilde{z}+z_r^{2\pi}(\tau)\right) - G\left(\tau,L\left(\tau,z_r^{2\pi}(\tau)\right),z_r^{2\pi}(\tau)\right) \tag{5.47}$$

$$\tilde{F}\left(\tau,x,\tilde{z}\right) \;\; = \;\; f\left(x,\alpha\left(x,\theta^* + \underbrace{\tilde{\theta} - \tilde{\theta}_r^{2\pi}(\tau)}_{\tilde{z}_1} + \tilde{\theta}_r^{2\pi}(\tau) + a\sin\tau\right)\right). \tag{5.48}$$

We note that

$$x = L\left(\tau,\tilde{z}+z_r^{2\pi}(\tau)\right) \tag{5.49}$$

is the *quasi-steady state*, and that the *reduced model*

$$\frac{d\tilde{z}_r}{d\tau} = \delta\tilde{G}\left(\tau,L\left(\tau,\tilde{z}_r+z_r^{2\pi}(\tau)\right),\tilde{z}_r+z_r^{2\pi}(\tau)\right) \tag{5.50}$$

has an equilibrium at the origin $\tilde{z}_r = 0$ (cf. (5.47) with (5.49)). This equilibrium has been shown in Section 5.3 to be exponentially stable for sufficiently small a.

To complete the singular perturbation analysis, we also study the *boundary layer model* (in the time scale $t = \tau/\omega$):

$$\begin{aligned}\frac{dx_b}{dt} &= \tilde{F}\left(\tau,x_b + L\left(\tau,\tilde{z}+z_r^{2\pi}(\tau)\right),\tilde{z}\right) \\ &= f\left(x_b + l(\theta),\alpha\left(x_b+l(\theta),\theta\right)\right),\end{aligned} \tag{5.51}$$

where $\theta = \theta^* + \tilde{\theta} + a\sin\tau$ should be viewed as a parameter independent from the time variable t. Since $f\left(l(\theta),\alpha\left(l(\theta),\theta\right)\right) \equiv 0$, then $x_b = 0$ is an equilibrium of (5.51). By Assumption 5.2, this equilibrium is exponentially stable.

By combining exponential stability of the reduced model (5.50) with the exponential stability of the boundary layer model (5.51), using Tikhonov's Theorem on the Infinite Interval [64, Theorem 9.4], we conclude the following:

- The solution $z(\tau)$ of (5.42) is $O(\omega)$-close to the solution $z_r(\tau)$ of (5.50), and therefore, it converges exponentially to an $O(\omega)$-neighborhood of the periodic solution $z_r^{2\pi}(\tau)$, which is $O(\delta)$-close to the equilibrium $z_r^{a,e}$. This, in turn, implies that the solution $\tilde{\theta}(\tau)$ of (5.18) converges exponentially to an $O(\omega+\delta)$-neighborhood of $-\dfrac{(h\circ l)'''(\theta^*)}{8\,(h\circ l)''(\theta^*)}a^2 + O\left(a^3\right)$. It follows then that $\theta(\tau) = \theta^* + \tilde{\theta}(\tau) + a\sin\tau$ converges exponentially to an $O(\omega+\delta+a)$-neighborhood of θ^*.

- The solution $x(\tau)$ of (5.46) (which is the same as (5.17)) satisfies

$$x(\tau) - l\left(\theta^* + \tilde{\theta}_r(\tau) + a\sin\tau\right) - x_b(t) = O(\omega), \tag{5.52}$$

where $\tilde{\theta}_r(\tau)$ is the solution of the reduced model (5.20) and $x_b(t)$ is the solution of the boundary layer model (5.51). From (5.52) we get

$$x(\tau) - l(\theta^*) = O(\omega) + l\left(\theta^* + \tilde{\theta}_r(\tau) + a\sin\omega\tau\right) - l(\theta^*) - x_b(t). \tag{5.53}$$

Since $\tilde{\theta}_r(\tau)$ converges exponentially to the periodic solution $\tilde{\theta}_r^{2\pi}(\tau)$, which is $O(\delta)$-close to the average equilibrium $\dfrac{(h \circ l)'''(\theta^*)}{8(h \circ l)''(\theta^*)}a^2 + O\left(a^3\right)$, and since the solution $x_b(t)$ of (5.51) is exponentially decaying, then by (5.53), $x(\tau) - l(\theta^*)$ exponentially converges to an $O(\omega + \delta + a)$-neighborhood of zero. Consequently, $y = h(x)$ converges exponentially to an $O(\omega + \delta + a)$-neighborhood of its maximal equilibrium value $h \circ l(\theta^*)$.

We summarize the above conclusions in the following theorem.

Theorem 5.5 *Consider the feedback system (5.11)–(5.14) under Assumptions 5.1–5.3. There exists a ball of initial conditions around the point $\left(x, \hat{\theta}, \xi, \eta\right) = (l(\theta^*), \theta^*, 0, h \circ l(\theta^*))$ and constants $\bar{\omega}, \bar{\delta}$, and $\bar{a}$ such that for all $\omega \in (0, \bar{\omega}), \delta \in (0, \bar{\delta})$, and $a \in (0, \bar{a})$, the solution $\left(x(t), \hat{\theta}(t), \xi(t), \eta(t)\right)$ converges exponentially to an $O(\omega + \delta + a)$-neighborhood of that point. Furthermore, $y(t)$ converges to an $O(\omega + \delta + a)$-neighborhood of $h \circ l(\theta^*)$.*

A considerably more elaborate analysis would lead to the following stronger result, which we give without proof.

Theorem 5.6 *Under the conditions of Theorem 5.5, there exists a unique exponentially stable periodic solution of (5.11)–(5.14) in an $O(\omega + \delta + a)$-neighborhood of the point $\left(x, \hat{\theta}, \xi, \eta\right) = (l(\theta^*), \theta^*, 0, h \circ l(\theta^*))$.*

Notes and References

This chapter is based upon [70], where the first proof of stability of extremum seeking appeared. The pioneering averaging studies of Meerkov [81, 82, 83] stand out as a precursor to the stability results presented in [70]. This analysis was instrumental in reviving interest in analysis and design for extremum seeking control. On a brief comparison of the results in this chapter with the results in Chapters 1, 2, we can infer that the filters of the extremum seeking scheme need not be very slow as was assumed here. Moreover, the low pass filter after demodulation is theoretically unnecessary, but useful in practice.

Chapter 6

Limit Cycle Minimization

Limit cycles occur in numerous areas of application. In particular, there are systems in which feedback control can only reduce the size of the limit cycle but cannot completely eliminate it. The inability to remove the limit cycle and achieve equilibrium stabilization may be associated with actuator constraints, such as magnitude and rate saturation. In this situation, the best control requirement is to enforce a stable, "smallest" limit cycle.

This chapter extends the extremum seeking scheme and its analysis in Chapter 5 to the case where equilibrium operation is not possible and the system is always in a *limit cycle*. The objective of the scheme is to reduce the size of the limit cycle to a minimum.

We start in Section 6.1 with a scheme for general feedback systems in limit cycle. This scheme incorporates a block for detection of the "amplitude" of the limit cycle. In Section 6.2 we apply the scheme to a Van der Pol oscillator example for which the simulations demonstrate the effectiveness of the scheme. Finally, in Section 6.3 we present stability/performance analysis which involves two steps of averaging with one step of singular perturbation analysis in between. The conclusions drawn are valid on $O(1)$ time intervals.

6.1 Scheme for Limit Cycle Minimization

We consider systems of the form

$$\begin{aligned}
\dot{x} &= f(x, u) \\
y &= h(x),
\end{aligned} \tag{6.1}$$

where $x \in \mathbb{R}^n$ is the state, $u \in \mathbb{R}$ is the input, $y \in \mathbb{R}$ is the output, and $f : \mathbb{R}^n \times \mathbb{R} \to \mathbb{R}^n$ and $h : \mathbb{R}^n \to \mathbb{R}$ are smooth. Suppose that we know a smooth control law

$$u = \alpha(x, \theta) \tag{6.2}$$

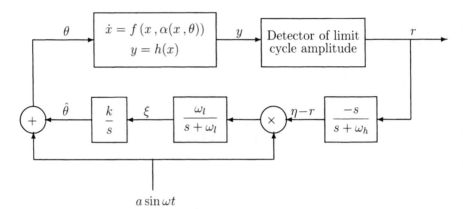

Figure 6.1: Extremum seeking scheme for limit cycle minimization. The x-system is assumed to be in a limit cycle for any constant θ (despite the use of feedback $\alpha(x,\theta)$).

parameterized by a scalar parameter θ such that the closed-loop system

$$\dot{x} = f(x,\alpha(x,\theta)) \tag{6.3}$$

has a stable limit cycle corresponding to each θ. Our objective is to tune θ to minimize the "amplitude" of the limit cycle.

Only a small modification is needed to adapt the extremum seeking scheme in Chapter 5 to the problem of limit cycle minimization. We add the detector block shown in Figure 6.2 to the overall extremum seeking scheme in Figure 6.1. The idea of the detector is simple. We assume that the output of the system in a limit cycle is sinusoidal, $y(t) = Y_0 + r\sin(\omega_0 t + \phi)$. The high pass filter is supposed to eliminate the DC component Y_0. The expected result, $r\sin(\omega_0 t + \phi)$, is squared to get $\frac{r^2}{2}(1 + \cos(2\omega_0 t + \phi))$, and then passed through a low pass filter to extract only $\frac{r^2}{2}$. The last block results in the amplitude of the limit cycle r. This idea is, of course, based on the assumption that $\omega_0 \gg \Omega_h, \Omega_l$.

The design parameters of the entire scheme are selected as $\omega_0 \gg \Omega_h, \Omega_l \gg \omega \gg \omega_h, \omega_l, k$.

6.2 Van der Pol Example

Consider a Van der Pol equation parameterized by θ as follows:

$$\ddot{x} + \epsilon\left[(x - x_0)^2 - 1 - (\theta - \theta^*)^2\right]\dot{x} + \mu^2(x - x_0) = 0, \tag{6.4}$$

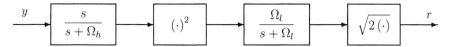

Figure 6.2: Detector of limit cycle amplitude

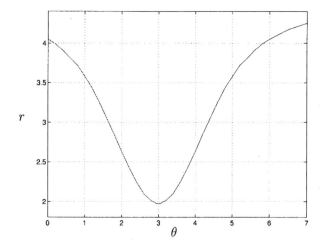

Figure 6.3: Characteristic of the limit cycle "amplitude" r with respect to θ.

where $\theta - \theta^*$ is a parameter that controls the amplitude of oscillation and x_0 is a parameter for the offset of x. We assume that θ^* is constant and θ is available as the input to the system. The system (6.4) will be in a limit cycle for any θ and θ^*. This example is contrived to emulate problems in which feedback control can reduce the size of a limit cycle but cannot eliminate it completely.

We first study the relationship between the limit cycle amplitude and the parameter θ for the system (6.4). The relationship is shown in Figure 6.3. Since the characteristic has a minimum, we feed $-r$ to the input of the extremum seeking block (see Figure 6.1).

We perform simulations from both sides of the extremum. In both cases, we set $\Omega_h = 0.75$, $\Omega_l = 0.02$, $\omega = 0.1$, $\omega_H = 0.02$, $k = 4$, $x_0 = 6$, $\theta^* = 3$, and $\epsilon = \mu = 1$. In the first case, we set the initial value of the integrator $\hat{\theta}(0) = 5$. We run the simulation without extremum seeking for 100 seconds and then start the extremum seeking controller. The oscillation of x is shown in Figure 6.4 and the process of convergence of the parameter θ to $\theta^* = 3$ is shown in Figure 6.5. In the second case, we consider the initial value $\hat{\theta}(0) = 1$. The oscillation of x is shown in Figure 6.6 and the process of convergence of θ to $\theta^* = 3$ is shown in Figure 6.7. In both cases the limit cycle is reduced to

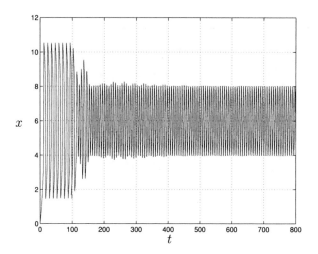

Figure 6.4: Time response of state x of the Van der Pol system with large $\hat{\theta}(0)$.

its minimal possible size.

6.3 Analysis

To simplify the analysis, we replace the amplitude detector block with a quadratic function. We also drop the low pass filter from the extremum seeking scheme to make the proof as simple as possible. The resulting extremum seeking scheme with the Van der Pol system is shown in Figure 6.8. Denoting $\tilde{\theta} = \theta - \theta^*$ and $y = x - x_0$, the system can be written as

$$0 = \ddot{y} - \epsilon \left(1 + (\tilde{\theta} + a \sin \omega t)^2 - y^2 \right) \dot{y} + \mu^2 y \tag{6.5}$$

$$\dot{\eta} = (y^2 - \eta)\omega_h \tag{6.6}$$

$$\dot{\hat{\theta}} = -ka(y^2 - \eta) \sin \omega t \tag{6.7}$$

To represent the Van der Pol system in polar coordinates, let

$$y = r \sin \phi, \qquad \dot{y} = \mu r \cos \phi. \tag{6.8}$$

Then, we have

$$\dot{r} = \epsilon r \cos^2 \phi \left[1 + \left(\tilde{\theta} + a \sin \omega t \right)^2 - r^2 \sin^2 \phi \right] \tag{6.9}$$

$$\dot{\phi} = \mu - \epsilon \cos \phi \sin \phi \left[1 + \left(\tilde{\theta} + a \sin \omega t \right)^2 - r^2 \sin^2 \phi \right]. \tag{6.10}$$

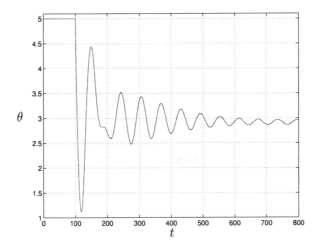

Figure 6.5: Time response of parameter θ of the Van der Pol system with large $\hat{\theta}(0)$.

The overall system is shown in (6.6)-(6.7) and (6.9)-(6.10). We treat t as a state and use ϕ as an independent variable. Then the whole system can be represented as

$$\frac{dr}{d\phi} = \frac{\epsilon}{\mu} \frac{r\cos^2\phi \left[1 + \left(\tilde{\theta} + a\sin\omega t\right)^2 - r^2\sin^2\phi\right]}{1 - \Delta} \tag{6.11}$$

$$\frac{d\tilde{\theta}}{d\phi} = -\frac{k}{\mu} \frac{a(r^2\sin^2\phi - \eta)\sin\omega t}{1 - \Delta} \tag{6.12}$$

$$\frac{d\eta}{d\phi} = \frac{\omega_h}{\mu} \frac{r^2\sin^2\phi - \eta}{1 - \Delta} \tag{6.13}$$

$$\frac{dt}{d\phi} = \frac{1}{\mu} \frac{1}{1 - \Delta}, \tag{6.14}$$

where

$$\Delta \triangleq \frac{\epsilon}{\mu}\cos\phi\sin\phi \left[1 + \left(\tilde{\theta} + a\sin\omega t\right)^2 - r^2\sin^2\phi\right]. \tag{6.15}$$

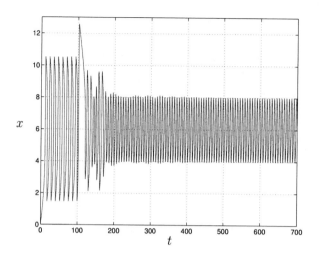

Figure 6.6: Time response of state x of the Van der Pol system with small $\hat{\theta}(0)$.

Now averaging with respect to ϕ for $\dfrac{1}{\mu}$ small, we obtain

$$\frac{dr^a}{d\phi} = \frac{\epsilon}{\mu} r^a \left[\frac{1 + (\tilde{\theta}^a + a\sin\omega t^a)^2}{2} - \frac{(r^a)^2}{8} \right] \tag{6.16}$$

$$\frac{d\tilde{\theta}^a}{d\phi} = -\frac{k}{\mu} a\sin\omega t^a \left(\frac{(r^a)^2}{2} - \eta^a \right) \tag{6.17}$$

$$\frac{d\eta^a}{d\phi} = \frac{\omega_h}{\mu} \left(\frac{(r^a)^2}{2} - \eta^a \right) \tag{6.18}$$

$$\frac{dt^a}{d\phi} = \frac{1}{\mu} \tag{6.19}$$

Note that $t^a = \dfrac{\phi}{\mu}$ in this average system. Denote $\phi_\tau = \dfrac{\omega\phi}{\mu}$. By using the relationship $\omega \gg \omega_h$, k, (6.16)–(6.18) can be expressed as

$$\omega \frac{dr^a}{d\phi_\tau} = \epsilon\, r^a \left[\frac{1 + (\tilde{\theta}^a + a\sin\phi_\tau)^2}{2} - \frac{(r^a)^2}{8} \right] \tag{6.20}$$

$$\frac{d\tilde{\theta}^a}{d\phi_\tau} = -\frac{k}{\omega} a\sin\phi_\tau \left(\frac{(r^a)^2}{2} - \eta^a \right) \tag{6.21}$$

$$\frac{d\eta^a}{d\phi_\tau} = -\frac{\omega_h}{\omega} \left(\frac{(r^a)^2}{2} - \eta^a \right) \tag{6.22}$$

This system is in the standard singular perturbation form.

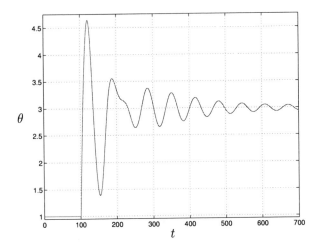

Figure 6.7: Time response of parameter θ of the Van der Pol system with small $\hat{\theta}(0)$.

The next step in our analysis is to study the system (6.20)–(6.22). We freeze r^a in (6.20) at its "quasi-steady state" value

$$(r^a)^2 = 4\left[1 + (\tilde{\theta}^a + a \sin \phi_\tau)^2\right] . \tag{6.23}$$

Substituting (6.23) into (6.21) and (6.22), we obtain the "reduced model"

$$\frac{d\tilde{\theta}_r^a}{d\phi_\tau} = -\frac{k}{\omega} a \sin \phi_\tau \left[2 + 2(\tilde{\theta}_r^a + a \sin \phi_\tau)^2 - \eta_r^a\right] \tag{6.24}$$

$$\frac{d\eta_r^a}{d\phi_\tau} = \frac{\omega_h}{\omega}\left[2 + 2(\tilde{\theta}_r^a + a \sin \phi_\tau)^2 - \eta_r^a\right] . \tag{6.25}$$

Since $\omega_h, k \ll \omega$, the system (6.24) and (6.25) is in the form to which the averaging method is applicable. The average model of (6.24) and (6.25) is

$$\frac{d\tilde{\theta}_r^{aa}}{d\phi_\tau} = -2\frac{k}{2\pi\omega} a \int_0^{2\pi} \sin \phi_\tau \left[2 + 2(\tilde{\theta}_r^{aa} + a \sin \phi_\tau)^2 - \eta_r^{aa}\right] d\phi_\tau \tag{6.26}$$

$$\frac{d\eta_r^{aa}}{d\phi_\tau} = \frac{\omega_h}{2\pi\omega} \int_0^{2\pi} \left[2 + 2(\tilde{\theta}_r^{aa} + a \sin \phi_\tau)^2 - \eta_r^{aa}\right] d\phi_\tau . \tag{6.27}$$

Performing the integrations, the average system becomes

$$\frac{d\tilde{\theta}_r^{aa}}{d\phi_\tau} = -2\frac{ka^2}{\omega} \tilde{\theta}_r^{aa} \tag{6.28}$$

$$\frac{d\eta_r^{aa}}{d\phi_\tau} = \frac{\omega_h}{\omega}\left[-\left(\eta_r^{aa} - 2 - a^2\right) + 2\left(\tilde{\theta}_r^{aa}\right)^2\right] \tag{6.29}$$

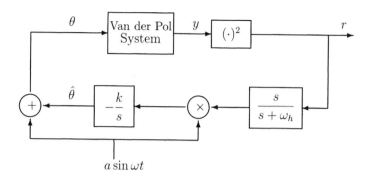

Figure 6.8: Simplified extremum seeking scheme for limit cycle minimization.

Define

$$\tilde{\eta}_r^{aa} = \eta_r^{aa} - (2 + a^2).$$
(6.30)

Then the average system is

$$\frac{d\tilde{\theta}_r^{aa}}{d\phi_\tau} = -2\frac{ka^2}{\omega}\tilde{\theta}_r^{aa}$$
(6.31)

$$\frac{d\tilde{\eta}_r^{aa}}{d\phi_\tau} = \frac{\omega_h}{\omega}\left[-\tilde{\eta}_r^{aa} + 2\left(\tilde{\theta}_r^{aa}\right)^2\right]$$
(6.32)

The Jacobian at the average equilibrium $\tilde{\theta}_r^{aa} = \tilde{\eta}_r^{aa} = 0$ is

$$J_r^{aa} = \begin{bmatrix} -2\dfrac{k}{\omega}a^2 & 0 \\ 0 & -\dfrac{\omega_h}{\omega} \end{bmatrix}.$$
(6.33)

Obviously, J_r^{aa} is Hurwitz. This implies that the average equilibrium is exponentially stable. Then, according to the averaging theorem [64, Theorem 8.3] all solutions $(\tilde{\theta}_r^a(\phi_\tau), \tilde{\eta}_r^a(\phi_\tau))$ converge exponentially to an $O(\delta)$-neighborhood of the origin where

$$\delta = \frac{\max\{k, \omega_h\}}{\omega}.$$
(6.34)

Since (6.24)–(6.25) is the reduced model of the singularly perturbed system (6.20)–(6.22), by the Tikhonov-type theorem on the infinite interval [64, Theorem 9.4], we have that

$$\tilde{\theta}^a(\phi_\tau) - \tilde{\theta}_r^a(\phi_\tau) = O(\omega)$$
(6.35)

$$(r^a(\phi_\tau))^2 - 4\left[1 + \left(\tilde{\theta}_r^a(\phi_\tau) + a\sin\phi_\tau\right)^2\right] \xrightarrow{\text{exp.}} O\left(\omega^2\right)$$
(6.36)

because it is easy to verify that the boundary layer model

$$\frac{dr_b}{dt} = \epsilon \left(r_b + 2\sqrt{1 + \theta^2} \right) \left(\frac{1 + \theta^2}{2} - \frac{\left(r_b + 2\sqrt{1 + \theta^2} \right)^2}{8} \right) \tag{6.37}$$

has an exponentially stable equilibrium at $r_b = 0$ for all θ. The above conclusions imply that

$$\tilde{\theta}^a(\phi_\tau) \xrightarrow{\text{exp.}} O(\delta + \omega) \tag{6.38}$$

$$(r^a(\phi_\tau))^2 \xrightarrow{\text{exp.}} 4 + O\left(a^2 + \delta^2 + \omega^2 \right). \tag{6.39}$$

Since (6.20)–(6.22) is the average system of (6.11)–(6.14), from the averaging theorem it follows that

$$\tilde{\theta}(\phi) \longrightarrow O\left(\delta + \omega + \frac{1}{\mu} \right) \tag{6.40}$$

$$r(\phi) \longrightarrow 2 + O\left(a + \delta + \omega + \frac{1}{\mu} \right) \tag{6.41}$$

(at least on an $O(\mu)$ interval for ϕ). By an argument similar to that in [64, Theorem 8.4], we establish the same properties for $\tilde{\theta}$ and r as functions of time, i.e.,

$$\tilde{\theta}(t) \longrightarrow O\left(\delta + \omega + \frac{1}{\mu} \right) \tag{6.42}$$

$$r(t) \longrightarrow 2 + O\left(a + \delta + \omega + \frac{1}{\mu} \right) \tag{6.43}$$

(at least on an $O(1)$ interval for t). This, in turn, implies that

$$\sqrt{y(t)^2 + \frac{\dot{y}(t)^2}{\mu^2}} \longrightarrow 2 + O\left(a + \delta + \omega + \frac{1}{\mu} \right). \tag{6.44}$$

The last statement means that extremum seeking brings the limit cycle amplitude to within $O\left(a + \delta + \omega + \frac{1}{\mu} \right)$ of its minimum.

Notes and References

This chapter is based on the results in [114]. Combustion instability control (Chapter 10) is an application of the idea developed in this chapter.

Part II

APPLICATIONS

Chapter 7

Antilock Braking

To ease the reader into applications of extremum seeking, we start Part II of the book with a simple example, which is not a research result, but an illustration of the theoretical ideas from Part I. Antilock braking systems (ABS) are currently an important tool in automotive vehicles. They stop faster and make safer turns when their wheels are prevented from locking. When the wheels are locked, they start slipping and steering becomes impossible, leading to loss of control of the vehicle.

ABS design was proposed to deal with braking on a slippery surface, i.e., to prevent the wheels from locking and skidding. In principle, the idea of an ABS is: by measuring a wheel's angular velocity and possibly linear acceleration, a decision is made if the wheel is about to lock. If it is, the pressure in the brake cylinder has to be reduced until the angular velocity of the wheel exceeds some threshold value. At this time the pressure is allowed to increase again. In this way, the wheels are prevented from locking. This process basically mimics the braking action of an experienced human driver. Due to nonlinearity and uncertainty in the braking process, the design of an ABS is difficult.

The characteristic of the friction force acting on the tires has a maximum for a low (nonzero) wheel slip and decreases as the slip increases. Standard ABS systems apply braking pressure in a rapid intermittent fashion. In some of them, the purpose of the intermittent action is to "seek" the maximum of the friction characteristic. In this chapter, we apply the extremum seeking feedback scheme of Chapter 5 to an ABS.

7.1   Model of a Slipping Wheel

We consider only one wheel, (the 'unicycle' model) depicted in Figure 7.1. The tire dynamics are described by the following two equations

$$m\dot{u} = -N\mu(\lambda) \tag{7.1}$$
$$I\dot{\Omega} = -B\Omega + NR\mu(\lambda) - \tau_{\mathrm{B}}, \tag{7.2}$$

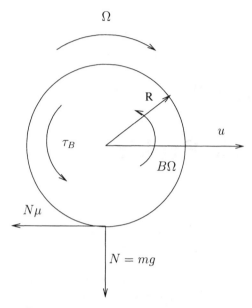

Figure 7.1: The wheel forces.

where u is the linear velocity and Ω the angular velocity of the wheel, m the mass, and $N = mg$ the weight of the wheel, R the radius of the wheel, I the moment of inertia of the wheel, $B\Omega$ the bearing friction torque, τ_B the breaking torque, $\mu(\lambda)$ the friction force coefficient, and the wheel *slip* λ is defined as

$$\lambda(u, \Omega) = \frac{u - R\Omega}{u} \qquad (7.3)$$

for the case of braking when $R\Omega \leq u$. The friction force coefficient $\mu(\lambda)$ is shown in Figure 7.2, from which it is seen that there exists an optimum μ^* at λ^*.

To formulate our problem into the extremum seeking setting, let us introduce a constant (which is unknown) λ_0 and define $\tilde{\lambda} = \lambda - \lambda_0$. The governing equation for $\tilde{\lambda}$ is:

$$\dot{\tilde{\lambda}} = \dot{\lambda} = \left(\frac{R\Omega}{u^2} + \frac{mR^2}{Iu} \right) \dot{u} + \frac{RB}{Iu}\Omega + \frac{R}{Iu}\tau_B . \qquad (7.4)$$

Since $\dot{u}$ is measurable via an accelerometer (they are also in use for airbags), it is easy to see that the simple feedback linearizing controller

$$\tau_B = -\frac{cIu}{R}(\lambda - \lambda_0) - B\Omega - \frac{I\Omega}{u}\dot{u} - mR\dot{u} , \qquad (7.5)$$

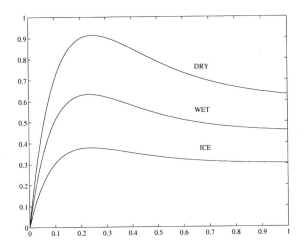

Figure 7.2: Friction force coefficient $\mu(\lambda)$.

where c is a positive constant, makes the equilibrium λ_0 of the system (7.4) exponentially stable, giving $\dot{\tilde{\lambda}} = -c\tilde{\lambda}$. Note that in the control τ_B we do not require knowledge of the unknown function $\mu(\lambda)$.

To maximize the friction force μN is to maximize $\mu(\lambda)$. Thus we can define the output for the system (7.4) as $y = \mu(\lambda)$ and maximize y. If $\mu(\lambda)$ is known, λ_0 can chosen to be at the optimum point λ^*, then λ will converge, exponentially, to the optimum value and the maximum friction force will be reached. However, λ_0 could not be exactly chosen at the optimum point because $\mu(\lambda)$ is not known and, as we can see from Figure 7.2, the optimum value differs for different road conditions. In the next section, we employ the extremum seeking scheme to search for the optimal value of λ in the face of uncertain road conditions.

7.2 ABS via Extremum Seeking

Fitting into the nomenclature of Section 1.2, the wheel model under feedback Eqn. (7.5) can be written as a cascade of input dynamics and a static map:

$$\frac{1}{c}\dot{\lambda} = -\lambda + \lambda_0 \tag{7.6}$$

$$y = \mu(\lambda). \tag{7.7}$$

We apply the scheme given in Figure 7.3 with

$$\lambda_0 = \hat{\lambda}_0 + a\sin\omega t. \tag{7.8}$$

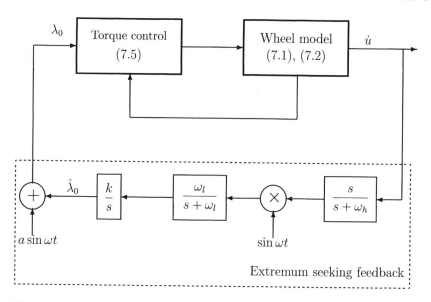

Figure 7.3: Extremum seeking scheme for the wheel model (7.1), (7.2).

For simulations, we use a simple function which qualitatively matches $\mu(\lambda)$ in Figure 7.2:

$$\mu(\lambda) = 2\mu^* \frac{\lambda^* \lambda}{\lambda^{*2} + \lambda^2}. \tag{7.9}$$

This function has a maximum at $\lambda = \lambda^*$, whose value is $\mu(\lambda^*) = \mu^*$. We run the test for $\lambda^* = 0.25$ and $\mu^* = 0.6$. As we mentioned before, neither the controller τ_B nor the extremum seeking scheme use the knowledge of the function $\mu(\lambda)$.

The wheel/quarter car parameters are chosen as: $m = 400kg$, $B = 0.01$, $R = 0.3m$. Initial conditions are: linear velocity, $u(0) = 120km/hr = 33.33m/s$; angular velocity, $w(0) = 400/3.6$, which makes $\lambda(0) = 0$.

Our simulation employs the extremum seeking scheme with $a = 0.01$, $\omega = 3$, $\omega_h = 0.6$, $\omega_l = 0.8$, and $k = 1.5$. For $\lambda_0(0) = 0.1$, the simulation results are shown in Figure 7.4. It is seen that during braking, maximum friction force is reached and the car is stopped within the shortest time and distance. The low pass filter in this design can be removed without loss of stability, i.e., $\omega_l = 0$. Its purpose is to attenuate noise in the loop.

Notes and References

This chapter is based on Chapter 9 of [74]. Several results exist where sliding-mode control was used [28, 35, 106, 110] to search for the optimum. While sliding mode control may cause chattering, the oscillations used in extremum

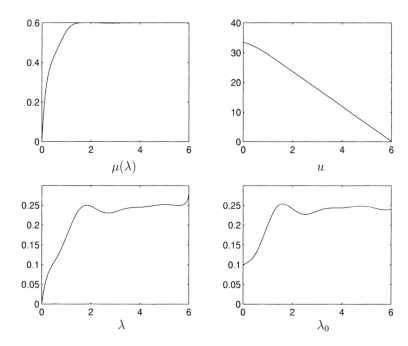

Figure 7.4: Antilock braking via extremum seeking.

seeking are much slower than those that noise would induce in sliding mode control.

Chapter 8

Bioreactors

Although not as complex as applications in subsequent chapters, the application in this chapter is more challenging than the ABS application in Chapter 7. The use of modern model-based techniques for optimization and control of bioreactors is hampered by a major bottleneck: the difficulty of identifying reliable first principle models for these highly nonlinear and widely uncertain systems.

It is, however recognized that even small performance improvements may result in substantial economic benefits. This chapter presents an "extremum seeking" approach for the optimization of bioreactors which allows automated seeking of the best operating point while being robust against a large uncertainty regarding the process kinetics. It uses the results of Chapter 5 for general nonlinear plants.

In this chapter, we maximize the productivity of a *continuous stirred tank bioreactor*. Compared to classical adaptive and neural net methods, the advantages of our approach are twofold: first the optimization objective (productivity maximization) is an explicit ingredient of the formulation of the adaptive control law, i.e., the optimization objective is guaranteed to be achieved when the control is convergent; second, this approach does not require any parametrization nor structural formalization of the modeling uncertainty (even under the form of a black box model like neural nets).

As a benchmark for our demonstration, we use a simple model of a continuous stirred tank biological reactor with numerical parameter values from [34, 54]. The optimization objective is to maximize the biomass production, more precisely the mass outflow rate of produced microorganisms. The steady states of the process can be characterized by a non-monotonic map relating the biomass production to the dilution rate which is our control input. The purpose of the extremum seeking method is to iteratively adjust the dilution rate in order to steer the process to the maximum of the map which corresponds to a maximum productivity.

This chapter is organized as follows. In Section 8.1 we describe the dynam-

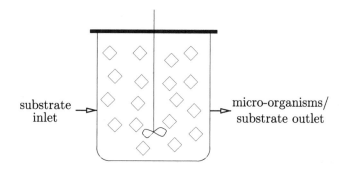

Figure 8.1: Bioreactor with continuous culture

ical model of the bioreactor under consideration, with Monod and Haldane kinetics and in Section 8.2 we state the control objective. In Section 8.3, we study the open loop stability of these models. We apply extremum seeking to the system in Section 8.4 and show simulation results. In Section 8.5 we design a stabilizing controller with a washout filter to extend the operating range for the Haldane model.

8.1 Dynamic Model of a Continuous Stirred Tank Reactor

In this section, we present the dynamic model of a continuous stirred tank bioreactor where a single population of micro-organisms is cultivated on a single limiting substrate. The bioreactor is shown in Figure 8.1.

The limiting substrate is fed into the culture vessel with a constant concentration s_R at a volumetric flow rate f. The culture medium is withdrawn at the same volumetric rate f so that the culture volume v in the vessel is kept constant. The dilution rate D is defined as $D = \dfrac{f}{v}$ and is the inverse of the residence time.

It is assumed that the other required substrates (including oxygen if needed) are provided in excess, that the culture medium is perfectly mixed and that the environmental conditions (temperature and pH) are regulated at appropriate constant values.

The dynamic behavior of this bioreactor is then described by the following standard mass balance model (see e.g. [17]):

$$\dot{x} = \mu(s)x - Dx \tag{8.1}$$

$$\dot{s} = D(s_R - s) - \frac{\mu(s)x}{Y} \tag{8.2}$$

where x is the biomass concentration, s the substrate concentration, $\mu(s)$ the specific growth rate function, and Y the yield coefficient.

Many analytical expressions for the function $\mu(s)$ have been proposed empirically or experimentally and we consider the two most commonly used, but many others could be considered as well. The most classical function is the Monod model:

$$\mu(s) = \mu_m \left(\frac{s}{K_s + s} \right) \tag{8.3}$$

where μ_m is the maximum growth rate constant and K_s is a saturation constant. If substrate inhibition is considered, the function $\mu(s)$ may be given by the Haldane model:

$$\mu(s) = \frac{\mu_m}{1 + \dfrac{K_s}{s} + \dfrac{s}{K_i}} \tag{8.4}$$

where K_i is an inhibition constant.

Hence, for Monod kinetics the bioreactor model is:

$$\dot{x} = x \left(\frac{\mu_m s}{K_s + s} - D \right) \tag{8.5}$$

$$\dot{s} = D(s_R - s) - \frac{\mu_m}{Y} \left(\frac{xs}{K_s + s} \right) \tag{8.6}$$

and for Haldane kinetics the model is:

$$\dot{x} = x \left(\frac{\mu_m}{1 + \dfrac{K_s}{s} + \dfrac{s}{K_i}} - D \right) \tag{8.7}$$

$$\dot{s} = D(s_R - s) - \frac{\mu_m}{Y} \frac{x}{1 + \dfrac{K_s}{s} + \dfrac{s}{K_i}} \cdot \tag{8.8}$$

To normalize the model, we use $Y s_R, s_R, \mu_m, \dfrac{1}{\mu_m}$ as the units of x, s, D, and t, respectively. So the nondimensional models become

$$\dot{x} = x \left(\frac{s}{K_1 + s} - D \right) \tag{8.9}$$

$$\dot{s} = D(1 - s) - \frac{xs}{K_1 + s} \tag{8.10}$$

for the Monod model and

$$\dot{x} = x \left(\frac{1}{1 + \dfrac{K_1}{s} + \dfrac{s}{K_2}} - D \right) \tag{8.11}$$

$$\dot{s} = D(1 - s) - \frac{x}{1 + \dfrac{K_1}{s} + \dfrac{s}{K_2}} \tag{8.12}$$

for the Haldane model, where $K_1 = \dfrac{K_s}{s_R}$ and $K_2 = \dfrac{K_i}{s_R}$.

8.2 Optimization Objective

Let us assume that the industrial goal of the process is the production of micro-organisms. As an optimization objective, it is then natural to consider the maximization of the amount of biomass harvested per unit of time which can be measured by the biomass outflow rate:

$$y = xD. \tag{8.13}$$

We shall see in the next section that the steady states of the process are characterized by a non-monotonic map relating the biomass outflow rate (the controlled output) y to the dilution rate D which is our control input. The purpose of the extremum seeking method is then to iteratively adjust the dilution rate in order to steer the process to the maximum of this map.

It is important to understand that we do *not* assume that the function $\mu(s)$ is a priori known: the Monod and Haldane models presented above must be viewed as a theoretical benchmark to illustrate and analyze the efficiency of the extremum seeking approach. Our aim will be to show that the best operating point can be discovered by a extremum seeking algorithm which is completely "ignorant" of the form of the kinetics.

8.3 Bifurcation Analysis of the Open-Loop System

8.3.1 Monod Model

To investigate the stability of the open-loop system with a Monod model, we first calculate equilibria corresponding to a constant dilution rate $D = D_0$. Let the right-hand side of (8.9) and (8.10) be zero. After some calculations we obtain two equilibria; one is $(x_0 = 0, s_0 = 1)$ and the other can be expressed as a function of D_0 as follows

$$s_0 = \frac{K_1 D_0}{1 - D_0} \tag{8.14}$$

$$x_0 = \frac{1 - (1 + K_1)D_0}{1 - D_0}. \tag{8.15}$$

The equilibrium $(x_0 = 0, s_0 = 1)$ is called the *wash-out steady state* since the concentration of the micro-organism is reduced to zero.

The Jacobian of the system at (x_0, s_0) is

$$J = \begin{bmatrix} \dfrac{s_0}{K_1 + s_0} - D_0 & \dfrac{K_1 s_0}{(K_1 + s_0)^2} \\[3mm] -\dfrac{s_0}{K_1 + s_0} & -\dfrac{K_1 x_0}{(K_1 + s_0)^2} - D_0 \end{bmatrix}. \tag{8.16}$$

It is easy to show that

1. the wash-out equilibrium ($x_0 = 0$, $s_0 = 1$) is stable when $D_0 > \dfrac{1}{1 + K_1}$
   and unstable when $D_0 < \dfrac{1}{1 + K_1}$.

2. at the other equilibrium, the Jacobian can be written as

$$J = \begin{bmatrix} 0 & B - D_0 \\ -D_0 & -B \end{bmatrix} \tag{8.17}$$

where

$$B \triangleq \dfrac{(1 - D_0)^2}{K_1} + D_0^2. \tag{8.18}$$

This equilibrium is defined only for $D_0 < \dfrac{1}{1 + K_1}$ and is stable for all the values of D_0 for which it is defined.

The steady state output can be expressed as

$$y_0 = \dfrac{D_0 (1 - (1 + K_1) D_0)}{1 - D_0}. \tag{8.19}$$

To obtain the extremum value y_0^* of y_0 we differentiate (8.19) with respect to D_0 and get

$$D^* = 1 - \sqrt{\dfrac{K_1}{1 + K_1}} \tag{8.20}$$

$$s^* = \sqrt{K_1(1 + K_1)} - K_1 \tag{8.21}$$

$$x^* = 1 + K_1 - \sqrt{K_1(1 + K_1)}. \tag{8.22}$$

For bifurcation analysis we select the parameters provided by Herbert et al. [54] as: $\mu_m = 1\text{hr}^{-1}$, $Y = 0.5$, $K_s = 0.2g/l$, $s_R = 10g/l$. So $K_1 = K_s/s_R = 0.02$. By substituting K_1 into (8.20)–(8.22), we get

$$y_0^* = 0.754 \quad \text{for} \quad D^* = 0.860, s^* = 0.123, x^* = 0.877. \tag{8.23}$$

Since the second derivative of y_0 with respect to D_0 is negative, this point is a maximum. The bifurcation diagram parameterized by the dilution rate for the steady state output is shown in Figure 8.2 in which the solid line represents stable equilibria and the dashed line represents unstable equilibria.

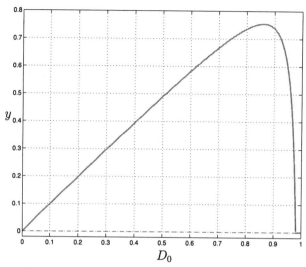

Figure 8.2: Bifurcation diagram of output of micro-organism w.r.t. dilution rate for a Monod model.

8.3.2   Haldane Model

To study the stability of the open-loop system with the Haldane model, we also calculate equilibria corresponding to a constant dilution rate $D = D_0$. Let the right-hand side of (8.11) and (8.12) be zero. Calculations show that the system has a unique equilibrium or multiple equilibria, depending on the value of D_0. The wash-out state always exists, i.e., $(x_0 = 0, s_0 = 1)$. For $D_0 < \dfrac{1}{1 + 2\sqrt{\frac{K_1}{K_2}}}$ there are two additional equilibria:

$$
\begin{cases}
x_{01} = 1 - s_{01} \\
s_{01} = \dfrac{K_2(1 - D_0)}{2D_0} - \dfrac{1}{2}\sqrt{\left(\dfrac{K_2(1 - D_0)}{D_0}\right)^2 - 4K_1K_2}
\end{cases}
\tag{8.24}
$$

$$
\begin{cases}
x_{02} = 1 - s_{02} \\
s_{02} = \dfrac{K_2(1 - D_0)}{2D_0} + \dfrac{1}{2}\sqrt{\left(\dfrac{K_2(1 - D_0)}{D_0}\right)^2 - 4K_1K_2}
\end{cases}
\tag{8.25}
$$

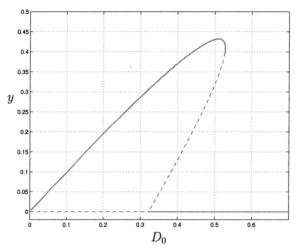

Figure 8.3: Bifurcation diagram of output of micro-organism w.r.t. dilution rate for a Haldane model.

The Jacobian at (x_0, s_0) is

$$
J = \begin{bmatrix}
\dfrac{1}{1 + \dfrac{K_1}{s_0} + \dfrac{s_0}{K_2}} - D_0 & -\dfrac{x_0\left(\dfrac{1}{K_2} - \dfrac{K_1}{s_0^2}\right)}{\left(1 + \dfrac{K_1}{s_0} + \dfrac{s_0}{K_2}\right)^2} \\[6mm]
-\dfrac{1}{1 + \dfrac{K_1}{s_0} + \dfrac{s_0}{K_2}} & -D_0 + \dfrac{x_0\left(\dfrac{1}{K_2} - \dfrac{K_1}{s_0^2}\right)}{\left(1 + \dfrac{K_1}{s_0} + \dfrac{s_0}{K_2}\right)^2}
\end{bmatrix}. \tag{8.26}
$$

It is easy to show that

1. the wash-out equilibrium $(x_0 = 0, s_0 = 1)$, is stable for $D_0 > \dfrac{1 + \frac{2}{K_2}}{\left(1 + K_1 + \frac{1}{K_2}\right)^2}$ and unstable for $D_0 < \dfrac{1 + \frac{2}{K_2}}{\left(1 + K_1 + \frac{1}{K_2}\right)^2}$.

2. at the other branch of equilibria, the Jacobian is

$$
J = \begin{bmatrix} 0 & -H \\ -D_0 & -D_0 + H \end{bmatrix} \tag{8.27}
$$

   where

$$
H \triangleq x_0 D_0^2 \left(\dfrac{1}{K_2} - \dfrac{K_1}{s_0^2}\right). \tag{8.28}
$$

We choose the parameters proposed by D'Ans and Kokotović [34]: $K_1 = 0.1$ and $K_2 = 0.5$. Substituting these values into the Jacobian *at the maximum point*, it becomes

$$J = \begin{bmatrix} 0 & 0.5099 \\ -0.5099 & -1.01977 \end{bmatrix}. \tag{8.29}$$

It is easy to check that the Jacobian is Hurwitz. A complete stability analysis along equilibria is shown in Figure 8.3.

The steady state output can be expressed as

$$y_0 = D_0 \left(1 - \frac{K_2(1 - D_0)}{2D_0} \pm \frac{1}{2} \sqrt{\left(\frac{K_2(1 - D_0)}{D_0} \right)^2 - 4K_1 K_2} \right). \tag{8.30}$$

To obtain the extremum value of y_0, we differentiate (8.30) with respect to D_0 and get

$$D^* = \frac{K_2 s^*}{(s^*)^2 + K_2 s^* + K_1 K_2} \tag{8.31}$$

$$s^* = \frac{\sqrt{K_1^2 K_2^2 + K_1 K_2^2 + K_1 K_2} - K_1 K_2}{1 + K_2} \tag{8.32}$$

$$x^* = 1 - \frac{\sqrt{K_1^2 K_2^2 + K_1 K_2^2 + K_1 K_2} - K_1 K_2}{1 + K_2}. \tag{8.33}$$

Substituting the values of K_1 and K_2 into (8.31)–(8.33), the maximum output is

$$y^* = 0.4322 \quad \text{for} \quad D^* = 0.5099 , s^* = 0.1523 , x^* = 0.8477 . \tag{8.34}$$

Since the second derivative is negative this point is a maximum. The bifurcation diagram for the output equilibrium parameterized by the dilution rate is shown in Figure 8.3.

8.4 Extremum Seeking via the Dilution Rate

Owing to the uncertainty and time-varying properties of biological processes the maximum operating point is hard to predict precisely. It is therefore of interest to implement extremum seeking control which is model-free and able to automatically tune the dilution rate in the right direction. A block diagram for extremum seeking implemented on a bioreactor is shown in Figure 8.4. The output performance index is the biomass outflow rate. The parameters are chosen as follows:

$$\text{speed of nonlinear dynamics} = O(1) \gg \omega \gg \omega_h , a , k . \tag{8.35}$$

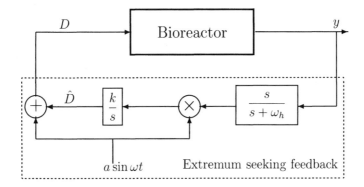

Figure 8.4: Extremum seeking scheme for a bioreactor.

We now demonstrate by simulations the ability of extremum seeking to adapt the dilution rate to optimize the biomass flow rate. We apply the (same) scheme to both the Monod and Haldane models.

8.4.1   Monod Model

For the Monod model, from Figure 8.2 we know that the peak occurs at $D^* = 0.86$, $s^* = 0.123$, $x^* = 0.877$. Our purpose is to tune D to D^*. We implement the extremum seeking scheme with the following choice of parameters:

$$\omega_h = 0.04 \,, \omega = 0.08 \,, a = 0.03 \,, k = 5 \,.$$

We start from an initial dilution rate lower than the optimum rate. Figure 8.5 shows how the extremum seeking approaches the peak along the equilibrium curve. The time response of the output is shown in Figure 8.6 and the time response of the tuning parameter is shown in Figure 8.7. The second simulation starts from a dilution rate larger than the optimum value. The results are shown in Figures 8.8–8.10.

From Figure 8.6, the settling time is 272 hours and the improvement in performance to the maximum output is 26.7%, which means the performance is improved with a rate of about 0.1%/hr. This rate of improvement is satisfactory but it is certainly not impressive. Since the time constants of the system at the peak are on the order of 10, this means that the convergence to the peak takes about 27 time constants. The convergence to the peak can be made faster by tuning the parameters of the scheme and by introducing an appropriate phase shift in the perturbation sinusoid. However, we do not do this here for two reasons. First, our primary objective is to qualitatively demonstrate the possibility of finding the peak, and not to optimize the transients. Second, and more important, if we chose parameters which make the

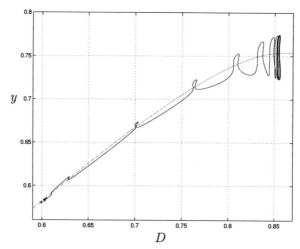

Figure 8.5: Maximum seeking process for the Monod model with initial dilution rate $D_0 = 0.6$.

convergence from the left side of the peak faster, they are too aggressive for the right side of the peak and may lead to instability. As evident by comparing Figures 8.6 and 8.9, the same parameters which result in relatively slow convergence from the left, result in fast convergence from the right. Since we do not assume to know the location of the peak, the adaptation must proceed cautiously. The oscillations of the output y in Figure 8.6 are about 3% of the peak equilibrium value of y.

8.4.2   Haldane Model

For the Haldane model, from Figure 8.3 we know that the peak occurs at $D^* = 0.5099$, $s^* = 0.1523$, $x^* = 0.8477$, $y^* = 0.4322$. Again our purpose is to tune D to D^*. We implement the extremum seeking scheme with the same parameters as those used with the Monod model.

We start from an initial dilution rate lower than the optimum value. Figure 8.11 shows how the extremum seeking approaches the peak along the equilibrium curve. The time response of the output is shown in Figure 8.12 and the time response of the tuning parameter is shown in Figure 8.13. If we increase the initial D_0 so that it is to the right of the optimum value, the time response of the output in Figure 8.14 shows that *the system falls to the washout steady state*. This is because the Haldane model has unstable equilibria underneath the maximum point. This motivates us to apply feedback control to stabilize the equilibrium branch under the maximum point, which is the subject of Section 8.5.

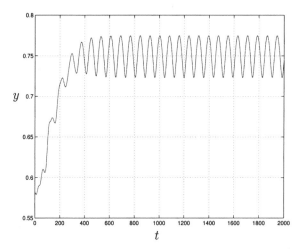

Figure 8.6: Time response of the output with extremum seeking for the Monod model with initial dilution rate $D_0 = 0.6$.

8.5 Feedback with Washout Filters for the Haldane Model

A washout filter

$$W(s) = \frac{s}{s + \omega_s} \qquad (8.36)$$

is a high pass filter that preserves the equilibrium structure and affects only the transient response and stability type.

In this section we assume the full state is measurable, and apply feedback

$$
\begin{aligned}
D &= D_0 + k_x(x - x_s) + k_s(s - s_s) & (8.37) \\
\dot{x}_s &= \omega_s x_s + \omega_s x & (8.38) \\
\dot{s}_s &= \omega_s s_s + \omega_s s \,. & (8.39)
\end{aligned}
$$

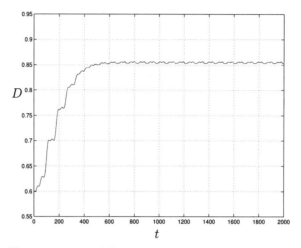

Figure 8.7: Time response of the tuning parameter with extremum seeking for the Monod model with initial dilution rate $D_0 = 0.6$.

8.5.1 Control Design

The Jacobian for the closed-loop system (x, s, x_s, s_s) at the equilibrium (x_0, s_0, x_0, s_0) is

$$
J = \begin{bmatrix}
-k_x x_0 & -\dfrac{x_0\left(\dfrac{1}{K_2} - \dfrac{K_1}{s_0^2}\right)}{\left(1 + \dfrac{K_1}{s_0} + \dfrac{s_0}{K_2}\right)^2} - k_s x_0 & k_x x_0 & k_s x_0 \\[4mm]
k_x x_0 - D_0 & -D_0 + k_s x_0 + \dfrac{x_0\left(\dfrac{1}{K_2} - \dfrac{K_1}{s_0^2}\right)}{\left(1 + \dfrac{K_1}{s_0} + \dfrac{s_0}{K_2}\right)^2} & -k_x x_0 & -k_s x_0 \\[4mm]
\omega_s & 0 & -\omega_s & 0 \\[2mm]
0 & \omega_s & 0 & -\omega_s
\end{bmatrix}.
$$

$$(8.40)$$

The eigenvalues of this fourth order matrix are difficult to calculate. However, we know that the eigenvalues are continuous. Therefore, for small ω_s, two of the eigenvalues will be approximately ω_s, and the other two will be approximately equal to the eigenvalues of the closed-loop system without washout filters. The characteristic polynomial at the peak is readily shown to be (using the values in D'Ans and Kokotović [34])

$$p(\lambda) = \lambda^2 + (0.8477(k_x - k_s) + 1.0198)\lambda + 0.4322(k_x - k_s) + 0.26. \quad (8.41)$$

By the Routh-Hurwitz method, the stability condition is

$$k_x - k_s + 1.20 > 0 \qquad \text{and} \qquad k_x + k_s < 0.60. \quad (8.42)$$

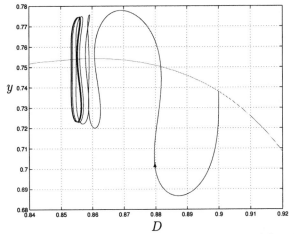

Figure 8.8: Maximum seeking process for the Monod model with initial dilution rate $D_0 = 0.9$.

By choosing $k_x = 0$ and $k_s = -0.2$, stability condition (8.42) is satisfied. The bifurcation diagram with D_0 as the parameter is shown in Figure 8.15. Note that a small amount of gain is sufficient to stabilize the entire branch of equilibria under the maximum point. We use this feedback gain for the extremum seeking simulation.

8.5.2   Simulation Results

As the starting point selected from Section 5.2, the initial dilution rate is 0.52. The parameters are selected as follows:

$$k_x = 0\,, k_s = -0.2\,, \omega_h = 0.04\,, \omega = 0.08\,, a = 0.03\,, k = 2\,, \text{ and } \omega_s = 0.01\,.$$

The seeking process is shown in Figure 8.16 and the time responses for the output y and the tuning parameter D are shown in Figures 8.17 and 8.18, respectively. Extremum seeking, combined with a small amount of stabilizing feedback, drives the system to the optimal equilibrium from a broad region of initial conditions. Thus, the stabilizing feedback improves the operating range of the system.

By applying extremum seeking to both Monod and Haldane models, we have shown that it can optimize the steady-state operation of a continuous stirred tank reactor in the face of uncertainty in the kinetics. The feedback is passed through a washout filter to keep the same structure of equilibria but only affect their stability type. As a result, the operating range of the system, and thus the region of applicability of extremum seeking, is extended.

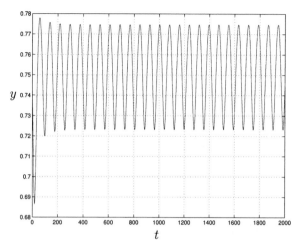

Figure 8.9: Time response of the output with extremum seeking for the Monod model with initial dilution rate $D_0 = 0.9$.

Notes and References

An extensive introduction to the modeling and control issues for bioreactors can be found in the tutorial paper [18]. For the feedback control of these processes, in order to cope with the modeling uncertainties, adaptive techniques have been mainly investigated in the literature (see, e.g., [17, 25, 111]) including more recently adaptive neural network models [24, 108, 121]. This chapter is based on [115]. In the Haldane model, a subcritical bifurcation prevents operation with a satisfactory stability region near the maximum of the biomass outflow rate. For this reason, a local stabilizing feedback is applied in [115] to soften the bifurcation. This provides a good candidate for optimization by the slope seeking method presented in Chapter 3. By doing so, we can achieve operation at a point below the maximum and will not need a locally stabilizing controller to stabilize the equilibria around the maximum (whose design needs some knowledge of the local dynamics).

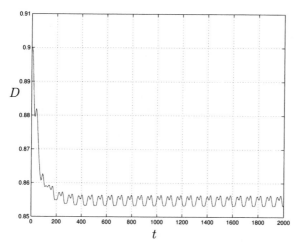

Figure 8.10: Time response of the tuning parameter with extremum seeking for the Monod model with initial dilution rate $D_0 = 0.9$.

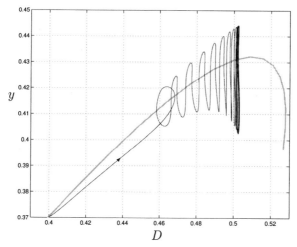

Figure 8.11: Maximum seeking process for the Haldane model with initial dilution rate $D_0 = 0.4$.

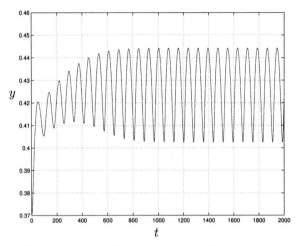

Figure 8.12: Time response of the output with extremum seeking for the Haldane model with initial dilution rate $D_0 = 0.4$.

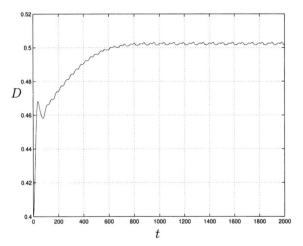

Figure 8.13: Time response of the tuning parameter with extremum seeking for the Haldane model with initial dilution rate $D_0 = 0.4$.

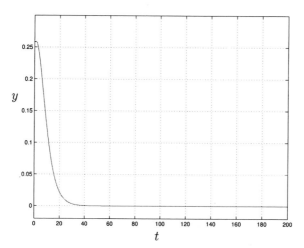

Figure 8.14: Time response of the output with extremum seeking for the Haldane model with initial dilution rate $D_0 = 0.52$. The system starts at an unstable equilibrium and *falls into the wash-out steady state*.

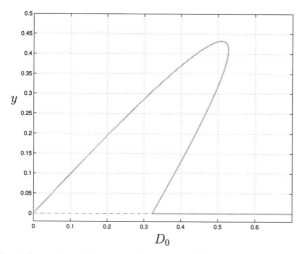

Figure 8.15: Bifurcation diagram of output of micro-organism w.r.t. dilution rate for a Haldane model with washout filters.

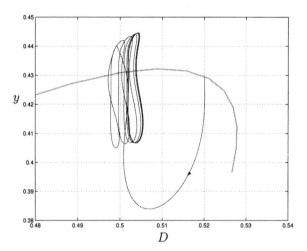

Figure 8.16: Maximum seeking process in for the Haldane model with state feedback and the initial dilution rate $D_0 = 0.52$.

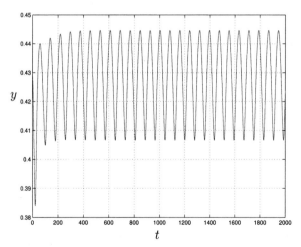

Figure 8.17: Extremum seeking time response of the output with state feedback for the Haldane model with initial dilution rate $D_0 = 0.52$.

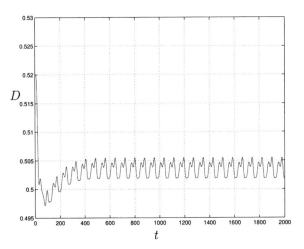

Figure 8.18: Extremum seeking time response of the tuning parameter with state feedback for the Haldane model with initial dilution rate $D_0 = 0.52$.

Chapter 9

Formation Flight

By flying in formation, similar to migratory birds [32], two or more aircraft can achieve a significant (up to 20% per aircraft) reduction in power demand [26, 56], which can be exploited to improve cruise performance, such as range and speed, or to increase the payload. This more efficient flight condition is attained through aerodynamic interference, by the wingman (the "follower" aircraft) riding upon the upwash field of the leader, like a glider in a thermal.

A distribution of wake velocity at two wingspans behind a C5-Galaxy is shown in Figure 9.1: the lateral distance y and vertical distance z from the aircraft are normalized by its wingspan b. This shows the existence an optimal configuration of the formation which yields maximal reduction in power demand. Roughly, this is a configuration where the wingman rides upon the peaks of the velocity distribution. This configuration can be reached and maintained with dedicated automatic control on the wingman. In fact, at the safe longitudinal separation of two wingspans, maintained between the aircraft (specifically, between the wing of the leader and the wing of the wingman) for collision avoidance [87], the effect of aerodynamic interference on the leader is marginal (weak dependence of formation flight benefits upon longitudinal separation permits freedom in setting it [56]) and, in any case, beneficial. Thus, in order to attain maximum-efficiency formation flight, only the wingman needs to be controlled, while the leader can be assumed to be stabilized in straight and level flight by an ordinary autopilot [27]. However, the sharp peaks of the velocity distribution in Figure 9.1 also indicate a high sensitivity of formation benefits to positioning error–this is shown in Figure 9.3. The figure plots relative range (relative to out of formation flight) versus lateral positioning error of the wingman from the optimum y_{opt} normalized by wingspan b. This brings out clearly the criticality of accurate tracking in this problem.

The wingman control system is based on a formation-hold autopilot (an autopilot capable of tracking relative position reference signals, i.e. wingman-leader separations signals), which is fed an estimate of the optimal separation. The estimate can be calculated from an aerodynamic interference model, or it

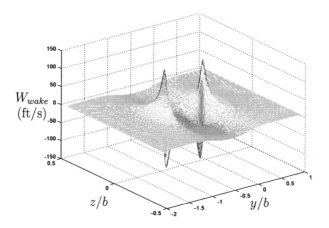

Figure 9.1: Distribution of wake velocity.

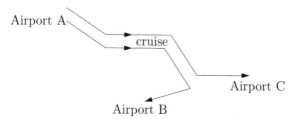

Figure 9.2: A cruise in formation.

can be generated by an adaptive feedback control scheme. Both these strategies have been adopted in studies on the problem, which has lately been a focus of intense interest, given the potential payoffs, and the availability of enabling avionics and control algorithms.

Here, we present *a generally applicable design procedure for minimum power demand formation flight with performance guarantees and easily measurable objective for extremum seeking.* This goal is attained through the following steps:

1. Modeling of aerodynamic interference as a multiple feedback nonlinearity in the aircraft dynamics.

2. The design of a new wake robust formation hold autopilot.

3. Transformation of the closed loop aircraft dynamics to a form in which the rigorous design procedure for extremum seeking from Chapter 2 is applicable.

4. Application of the design procedure from Chapter 2 to attain stable

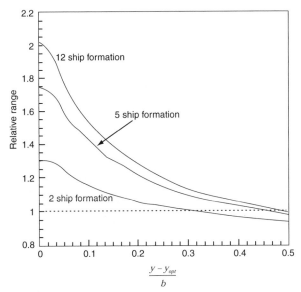

Figure 9.3: Dependence of formation benefits on positioning accuracy. (Reproduced from [22].)

extremum seeking minimizing the pitch angle of the wingman, an easily measurable objective, accounting for wake induced uncertainties.

We apply the design procedure on a formation of Lockheed C-5s, extending the use of maximum performance formation flight to large transports (the C-5's size is around that of the Boeing-747, and it is the largest military transport used by NATO). We use available experimental wake data of the C-5, to develop a model of the aircraft in the wake that models aerodynamic interference as feedback nonlinearities. Thus, this chapter presents stable extremum seeking for a plant with nonlinear feedback. The choice of the C-5 for study is motivated by the following: large transports flying long missions, mostly in a cruise condition, can get a high economical payoff from the system; the C-5 has a consistent fleet, which will stay in service for 40 more years with new avionics and engines [62, 95]; and experimental data on the wake of the C-5 is available [43].

The chapter is organized as follows. In Section 9.1 we model wingman dynamics in the wake of the leader. Section 9.2 details design of the new formation-hold autopilot. Section 9.3 provides the transformation of the optimal formation flight problem to the framework for extremum seeking design, and the extremum seeking design for the C-5s in formation. Section 9.4 presents simulations (all simulations in this chapter were performed in MATLAB and SIMULINK) of optimal formation flight of the C-5s in both calm air and in turbulent conditions.

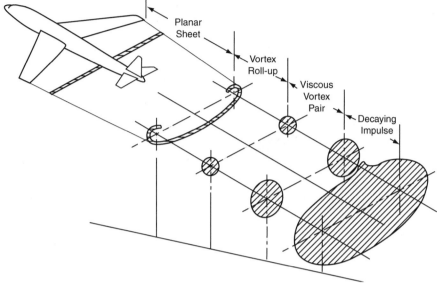

Figure 9.4: Wake vortex roll-up and decay. (Reproduced from [49].)

9.1 Wingman Dynamics in Close Formation Flight

The dynamics of an aircraft in close formation flight is much more complex compared to the dynamics in free flight because of aerodynamic interference (wake-induced forces and moments, which mean new terms in the equations) that arises from the wake generated by other aircraft. Since this formation flight phenomenon significantly alters wingman dynamics, its effects have to be sufficiently captured in modeling for control design, to ensure reliable performance of the control system in the real operating environment.

We solve the problem in four steps. We first develop a model of the wake of the leading C-5 from available wake data [43], neglecting influence of the wingman "far behind". Based on this model, aerodynamic forces and moments on the wingman in the wake of the leader are computed. Then an equilibrium study for the wingman in the wake is performed, yielding powerful insight into the physics of close formation flight benefits. Finally, the dynamics of the wingman in the wake is derived from free flight dynamics.

9.1.1 Wake Model

The wake of an aircraft can be described by: vortex sheet generation and roll-up, rolled-up wake structure, vortex transport, and vortex decay [49] (Figure 9.4). Two simplifying assumptions can be made immediately, thanks to the

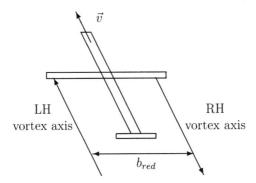

Figure 9.5: Aircraft trailing vortices.

value of the longitudinal separation between the aircraft. In fact, the safe two-wingspan figure is large enough for the roll-up to be complete, yet small enough to neglect the slow vortex decay process [99]. Vortex transport, which involves a change in orientation and a distortion of the vortex axes from their modeled configuration, is difficult or impossible to predict [49]. This phenomenon translates into uncertainties of the order of 20 ft [49] on the optimal separations of the wingman, enough to cut formation flight benefits by 50% [22]. Hence, we model the rolled-up wake structure as two counter-rotating semi-infinite straight vortices trailing from the wing [50], parallel to the flight path (the leader is assumed to be in straight and level flight all the time), and separated by a distance equal to the reduced wing span, b_{red}, as shown in Figure 9.5 and later design the adaptation to account for the position uncertainty due to vortex transport. For this purpose, the NASA-Burnham-Hallock tangential velocity profile [22] is used, because it correlates well with experimental data:

$$V_\theta(r) = \frac{\Gamma}{2\pi r}\frac{r^2}{(r_c^2 + r^2)}, \tag{9.1}$$

where V_θ is the tangential velocity, Γ is the circulation, r is the radial distance from the vortex axis, and r_c is the core radius of the vortex. The circulation is given by

$$\Gamma = \frac{W}{\rho V_\infty b_{red}}, \tag{9.2}$$

and the reduced wingspan by

$$b_{red} = \frac{\pi}{4}b, \tag{9.3}$$

where W is the aircraft weight, ρ is the air density, V_∞ is the reference airspeed, and b is the aircraft wingspan. Finally, based on experimental data [43], a 5 ft vortex core radius is estimated for the C-5 in the high altitude cruise flight condition. The wake-induced velocity field obtained with this model is given in Appendix C.

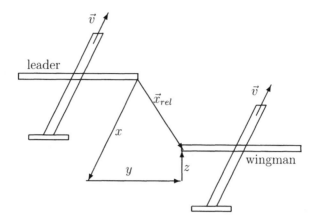

Figure 9.6: Configuration of formation flight.

9.1.2   Forces and Moments on the Wingman in the Wake

Using the wake-induced velocity field, forces and moments on the wingman in the wake of the leader are calculated, with an emphasis on simple modeling. This choice is crucial, because the alternative is to re-calculate the entire straight and level aerodynamic database of the C-5 for flight in the wake, a formidable task [23].

The closeness of aircraft conditions in the wake to trim conditions permits splitting the forces and moments on the aircraft into two terms: a free flight term, and an extra term due to flight in the wake. In this subsection we shall only be concerned with the latter, for which some powerful simplifications can be made. First of all, since the aircraft fly nearly straight and level and parallel, it can be assumed that extra forces and moments depend only on the relative position, i.e. separations, between the two aircraft, and on no other states. The relative position, $\vec{x}_{rel}$, and its components, the longitudinal, the lateral, the vertical separations, respectively, x, y, z, are defined in Figure 9.6. A second assumption can be made on the same grounds that allow decoupling between longitudinal and lateral-directional dynamics in free flight. It refers to specific forces and moments: the longitudinal and vertical forces, and the pitching moment depend only on the upwash distribution; the lateral force and the yawing moment are only due to the sidewash. No such simplification is possible for the rolling moment which depends on both the upwash and the sidewash distribution.

To determine the longitudinal and vertical forces, and the pitching moment, we assume that the effect of the upwash distribution is equivalent to that of a uniform distribution obtained by averaging the actual one along the wingspan [91]. Then we use the available stability derivatives to compute the forces and the moment in one shot. The average upwash, $\bar{W}_{wake}$, is given by

$$\bar{W}_{wake}(x, y, z) = \frac{1}{b} \int_0^b W_{wake}(x, y + s, z)c(s)ds, \qquad (9.4)$$

where b is the wingspan, W_{wake} is the upwash, $c(s)$ is the chord distribution, used as a weight for the average, and s is the lateral coordinate along the wingspan originating at the left wingtip.

The rolling moment due to the upwash, L_{wake}, is calculated using modified strip theory [49]:

$$L_{wake}(x, y, z) = -m\frac{1}{2}\rho_0 V_\infty a_0 \int_0^b W_{wake}(x, y + s, z)c(s)Q(s)sds, \qquad (9.5)$$

$$Q(s) = \frac{\pi}{4}\sqrt{1 - \left[\frac{2(s - b/2)}{b}\right]^2}, \qquad (9.6)$$

$$m = \frac{1}{1 + \left(\frac{2a_0}{\pi AR}\right)(1 + \varepsilon)}, \qquad \varepsilon = \frac{3TR - 1}{3(1 + TR)}, \qquad (9.7)$$

where $Q(s)$ is an elliptical weight, m is a correction factor, a_0 is the two-dimensional lift curve slope, AR is the aspect ratio, and TR is the taper ratio. A value of 5.67 is used for a_0, as recommended in [49]. The rolling moment is given in terms of the rolling moment factor R.M.F $= \frac{L_{wake}}{L_{max}}$.

Side-force and lateral-directional moments due to the sidewash were calculated assuming a uniform distribution, equal to the value of sidewash at the centerline, V_{wake}^{CL}. We use this simplification because sidewash-induced effects are small compared to other wake effects. The sign convention for V_{wake}^{CL}, is opposite to the one for V_{wake}.

With this simple modeling, both the longitudinal dynamics forces and moment, and the rolling moment are overestimated (although, a more careful choice of the weights can improve accuracy). This, however, shall not be a concern, because an overestimate of formation flight benefits leads to conservative design of the control system. More refined modeling, like vortex lattice [22, 79], can be used for fine tuning and analysis.

The average upwash, $\bar{W}_{wake}$, the rolling moment, L_{wake}, and the sidewash, V_{wake}^{CL}, fields are shown in Figs. 9.7, 9.8, and 9.9, as functions of the lateral separation, y, and of the vertical separation, z, at a longitudinal separation, $x = 2b$. In fact, there is no significant dependence upon the longitudinal separation: hence a constant value of two wingspans will be used for all aerodynamic interference calculations.

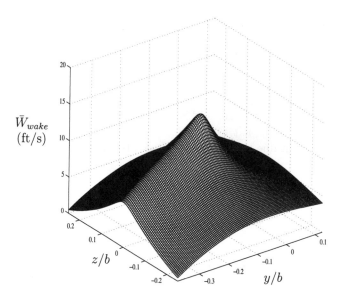

Figure 9.7: Average upwash.

9.1.3 Wingman Equilibrium in the Wake

A comparison of horizontal rectilinear flight of the wingman in and out of formation provides an estimate of formation flight benefits and links them to a measurable quantity, which is needed for the adaptive online optimization.

The use of the average upwash concept to model wake-induced forces in the vertical plane, permits proceeding in analogy with flight in uniform rising air. Conclusions can then be reached with simple application of small perturbation theory.

While in the wake, the wingman experiences the leader-induced upwash field, which translates into an increase of the angle of attack and thus of lift, unless speed is reduced at the same time. Then, in order to maintain vertical equilibrium at the same speed, the wingman has to pitch down. The more it pitches down, the more the weight helps thrust balance drag, as shown in Figure 9.10, where V_{air} is the airspeed, L is the lift, D is the drag, T is the thrust, and the subscript *form* refers to steady-state in formation. Hence, thrust reduction, i.e., formation flight benefits, are related to the average upwash and to the wingman steady-state pitch angle. The relationship is proportional and specifically:

$$\Delta T = T_{form} - T_0 \approx W \frac{\bar{W}_{wake}}{V_\infty}, \quad \bar{\theta} \approx -\delta \approx \frac{-57.3\,\bar{W}_{wake}}{V_\infty}\text{deg}, \qquad (9.8)$$

where the subscript 0 refers to steady-state out of formation. T_0 is 30,000 lb.

These conclusions have two important applications. First of all, they allow

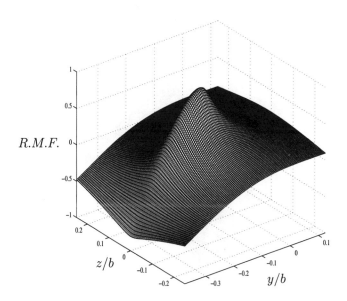

Figure 9.8: Rolling moment factor.

estimation of both the maximum thrust reduction and the relative position at which it is realized, by inspection of the average upwash plot, Figure 9.7:

$$\Delta T_{min} = -13000 \, \text{lb}, \quad \Delta T_{\%min} = \frac{\Delta T_{min}}{T_0} = -43\%, \quad \bar{\theta}_{min} = -1.13 \, \text{deg}, \quad (9.9)$$

$$y_{opt} = -24.64 \, \text{ft}, \quad z_{opt} = 0 \, \text{ft}, \quad (9.10)$$

where the subscript opt refers to the optimal configuration. Secondly, the pitch angle of the wingman, which can be easily measured, can be fed to the adaptive loop to achieve online optimization.

9.1.4   Wingman Dynamics in the Wake

The model is based on standard linearized decoupled dynamics in free flight, since state deviations from trim conditions are small. The reference condition chosen for design is cruise at Mach 0.77, 40000 ft, and 650,000 lb. Dynamics are then given by

$$\dot{\mathbf{x}} = A\mathbf{x} + B\mathbf{u}_c, \quad (9.11)$$

where $\mathbf{x} = \begin{pmatrix} \mathbf{x}_{long} & \mathbf{x}_{lat} \end{pmatrix}^T$, $\mathbf{u}_c = \begin{pmatrix} \mathbf{u}_{clong} & \mathbf{u}_{clat} \end{pmatrix}^T$ and $A = \text{diag}\begin{pmatrix} A_{long} & A_{lat} \end{pmatrix}$, $B = \text{diag}\begin{pmatrix} B_{long} & B_{lat} \end{pmatrix}$, and the subscript $long$ stands for longitudinal dynamics, and the subscript lat stands for lateral-directional dynamics. The states, $\mathbf{x}_{long}$, $\mathbf{x}_{lat}$, and the control inputs, $\mathbf{u}_{clong}$, $\mathbf{u}_{clat}$, are

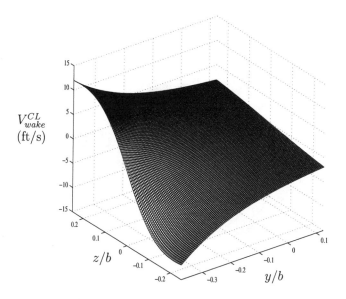

Figure 9.9: Sidewash at centerline.

given in Appendix C; stability derivatives are given in [53]. The dynamics and the saturation points of the conventional controls of the C-5 have been assumed, due to lack of data; they are given in Appendix C. The states can all be measured with accelerometers and gyros, coupled with DGPS (Differential Global Positioning System) and datalink between the two aircraft for separations [120]. From 2003, ring-laser gyros and DGPS will be standard equipment on the C-5. The measurement of angle of attack in the highly non-uniform wake-induced velocity field is not meaningful. Hence, its use in feedback should be avoided.

While perfectly adequate for free flight, linear modeling is not suitable for formation flight. Hence the dynamics of the wingman in the wake is derived from free flight dynamics, by incorporating formation-related extra forces and moments as feedback nonlinearities (see Figure 9.11):

$$\dot{\mathbf{x}} = A\mathbf{x} + B\mathbf{u}_c + F\mathbf{u}_{wake}(y, z), \tag{9.12}$$

where

$$\mathbf{u}_{wake}(y, z) = \left(\bar{W}_{wake}(y, z) \quad L_{wake}(y, z) \quad V_{wake}^{CL}(y, z) \right)^{T}, \tag{9.13}$$

and

$$F = \begin{pmatrix} F_W & 0 & 0 \\ 0 & F_L & F_V \end{pmatrix}. \tag{9.14}$$

The wake influence matrix, F, is given in Appendix C, through its three

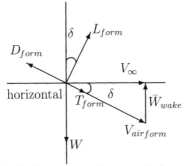

Figure 9.10: Forces on wingman in formation.

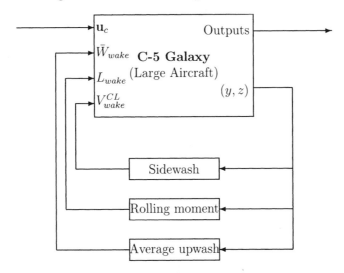

Figure 9.11: Wingman in the wake.

non-zero partitions, along with the units and the sign conventions for the wake-induced inputs, $\mathbf{u}_{wake}(y, z)$.

9.2   Formation-Hold Autopilot

The task of the formation-hold autopilot is to drive the wingman to the relative position (with respect to the leader in rectilinear flight) prescribed by the extremum seeking algorithm. This translates into the capability of tracking reference longitudinal, lateral and vertical separation signals.

The use of the autopilot in an adaptive loop with the purpose of maximum-efficiency flight in an uncertain wake-induced velocity field, produces unique design specifications:

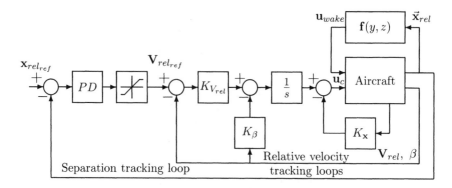

Figure 9.12: Formation-hold autopilot.

1. High speed tracking in a neighborhood of the optimal configuration to ensure closed loop stability and speed of convergence with adaptation.

2. Ability to track large reference position signals to enable formation join-in from afar.

3. Robustness of tracking performance to aerodynamic interference (This is crucial to extremum seeking design).

The uniqueness of these requirements dictate a new design approach, despite the availability of other formation-hold autopilots [45, 91, 101].

Architecture and Design.   The structure of the autopilot is shown in Figure 9.12. It uses full state measurement (available through an Inertial Navigation System, a DGPS, and a datalink between the aircraft). It consists of the following: a relative velocities tracking loop (which includes a turn coordination loop based on the sideslip angle β) designed in the error space with the internal model principle [41] with the feedback gain designed by linear quadratic regulation (LQR) (this is implemented through the two internal loops in Figure 9.12, one proportional to the state, the other proportional to the integral of relative velocities error); a separations tracking loop with classically designed proportional derivative (PD) compensators in the outer loop, along with rate limiters, placed between the inner and the outer loop. High speed tracking is attained by using high gains; actuator and engine saturation and integrator windup due to large join-in reference signals are prevented by the rate limiters. The following is a compact representation of the closed loop dynamics in Figure 9.12:

$$\dot{\mathbf{x}} = A\mathbf{x} + B[\frac{K_{\mathbf{V}_{rel}}}{s}\left[(PD\mathbf{r} - \mathbf{V}_{rel})\right] - K_{\mathbf{x}}\mathbf{x} - \frac{K_\beta}{s}\left[\beta\right]] + F\mathbf{u}_{wake}(y, z), \quad (9.15)$$

where

$$\mathbf{r} = \mathbf{x_{rel_{ref}}} - \vec{\mathbf{x}}_{rel}, \quad \vec{\mathbf{x}}_{rel} = (x \quad y \quad z)^T, \quad \mathbf{V_{rel}} = (V_x \quad V_y \quad V_z)^T \ (9.16)$$

and

$$K_{\mathbf{x}} = \mathrm{diag}\left(K_{\mathbf{x}_{long}} \quad K_{\mathbf{x}_{lat}}\right), \quad K_{\mathbf{V}_{rel}} = \begin{pmatrix} K_{V_x} & 0 & K_{V_z} \\ 0 & K_{V_y} & 0 \end{pmatrix}, \quad (9.17)$$

and

$$PD = \begin{pmatrix} k_{Px} + k_{Dx}s & 0 & 0 \\ 0 & k_{Py} + k_{Dy}s & 0 \\ 0 & 0 & k_{Pz} + k_{Dz}s \end{pmatrix}. \quad (9.18)$$

All autopilot parameters are supplied in Appendix C.5.

Robustness to Aerodynamic Interference.   This requires that the closed loop dynamics in Eqn. (9.15) be stable at all points in the wake. Linearizing Eqn. (9.15) about a point $(\bar{y}, \bar{z})$, in the wake, we get

$$\dot{\mathbf{x}} = A\mathbf{x} + B[\frac{K_{\mathbf{V}_{rel}}}{s}[(PD\mathbf{r} - \mathbf{V}_{rel})] - K_{\mathbf{x}}\mathbf{x} - \frac{K_\beta}{s}[\beta]] + F\frac{\partial \mathbf{u}_{wake}}{\partial \zeta}(\bar{y}, \bar{z})(\zeta - \bar{\zeta}),$$
$$(9.19)$$

where $\zeta = (y, z)$, and $\bar{\zeta} = (\bar{y}, \bar{z})$. The requirement of closed loop stability at all points $(\bar{y}, \bar{z})$ in the wake translates to stability of the system in Eqn. (9.19) for a range of gradients $\frac{\partial \mathbf{u}_{wake}}{\partial \zeta}(\bar{y}, \bar{z})$ of the feedback nonlinearity. The structure of Eqn. (9.19) has motivated design of high autopilot gains to achieve desired robustness of tracking performance to aerodynamic interference. The stability of the autopilot so designed has been checked by root locus calculations for the range of gradients expected in the wake. As a sample, Figure 9.13 shows the root locus (poles are shown with crosses and zeros with circles; some very large zeros in the transfer function are not shown) of the transfer function between average upwash $\bar{W}_{wake}$ and vertical separation z for the range of gradients in the average upwash field $[-0.345, +0.345](\mathrm{ft/s})/\mathrm{ft}$. Figure 9.14 presents a zoomed in version to show that the dominant poles hardly change resulting in performance practically identical to free flight operation. Robustness to aerodynamic interference is also illustrated by the simulation results below.

Simulation Results.   Figs. 9.15, 9.16 show a typical approach to the optimum assuming perfect knowledge of its position. The wingman is initialized 20 ft below and 20 ft to the right of the optimal position.

Two sets of time histories are shown for comparison: solid lines represent the autopilot performance with aerodynamic interference, while dashed lines represent autopilot performance without aerodynamic interference, which is the condition in which it has been designed. Vertical and lateral separation

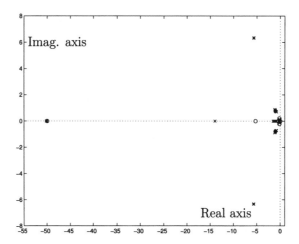

Figure 9.13: Autopilot root locus analysis: stability robustness in wake operation.

time histories exhibit almost identical performance with and without aerodynamic interference. The longitudinal separation error is not a concern, because it does not have any significant effect on formation flight benefits, and because it is well within safety margins for collision avoidance.

Elevator and ailerons deflections reach maximum values almost instantly, to guarantee maximum performance: longitudinal convergence time of 5 s, and lateral convergence time of 10 s. Vertical acceleration (not shown) does not exceed a peak of 0.3 g; lateral acceleration (not shown) is negligible, thanks to turn coordination. Other simulation runs (not shown here for space reasons) demonstrated good operation of the system starting at any distance from the leader, without reaching actuator saturation. Far away, the rate limiters set vertical approach speed at 500 ft/min and lateral approach speed at 250 ft/min.

As the optimum is reached, thrust is reduced by about 40%, as the aircraft pitches down by about one degree, consistently with the equilibrium analysis in section 9.1. The ailerons deflect by about 20 deg, compensating wake-induced rolling moment. Like thrust reduction, this is an overestimate due to the approximate calculation of aerodynamic interference; yet it suggests that high aileron trim drag is to be expected. A method to eliminate it is presented in the discussion in Section 9.4.

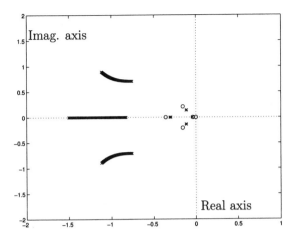

Figure 9.14: Autopilot root locus analysis: performance robustness in the wake.

9.3 Extremum Seeking Control of Formation Flight

The problem of minimizing power demand through formation flight appears to fit intuitively into the framework of extremum seeking control. The choice of objective used in this work (the pitch angle θ), however does not permit the problem to fit into the standard extremum seeking framework for which a rigorous design method is available. Hence, we transform our problem to fit into the standard framework, and then perform design. We, the transformation of our problem to fit the standard scheme of Chapter 2 in Subsection 9.3.1, and our design in Subsection 9.3.2.

9.3.1 Formulation as a Standard Extremum Seeking Problem

We show here that the extremum seeking scheme in Figure 9.17 can be transformed to the form in Figure 2.1, in which we can then use the available design algorithm. We achieve this objective through the following steps:

1. Write state space representations of the dynamics from the output of extremum seeking (y_{ref}, z_{ref}) to the relative position (y, z), and to the pitch angle θ:

$$\dot{\mathbf{x}}_{long} = \left(A_{long} - B_{long}K_{\mathbf{x}_{long}}\right)\mathbf{x}_{long}$$

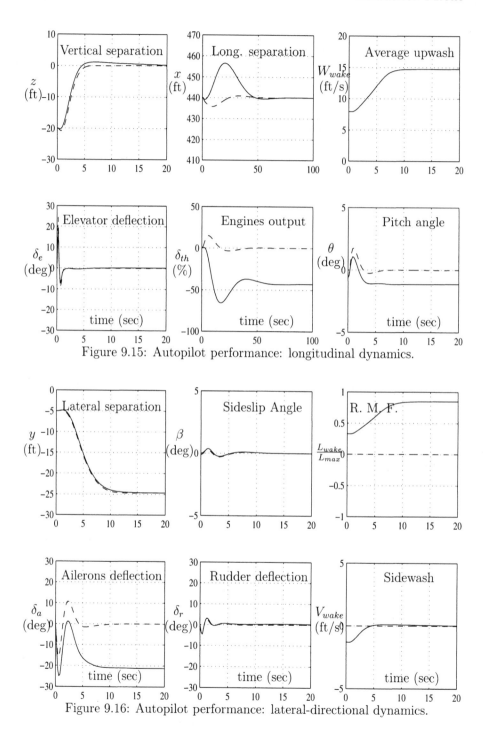

Figure 9.15: Autopilot performance: longitudinal dynamics.

Figure 9.16: Autopilot performance: lateral-directional dynamics.

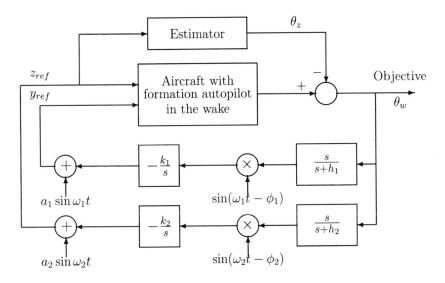

Figure 9.17: Extremum seeking for formation flight.

$$+B_{long}[\frac{K_{V_x}}{s}[(k_{Px} + k_{Dx}s)[x_{ref} - x] - V_x]]$$
$$+B_{long}[\frac{K_{V_z}}{s}[(k_{Pz} + k_{Dz}s)[z_{ref} - z] - V_z]] + F_W \bar{W}_{wake}(y, z)$$
$$z = C_z \mathbf{x}_{long}, \quad \theta = C_\theta \mathbf{x}_{long}. \tag{9.20}$$

and

$$\dot{\mathbf{x}}_{lat} = (A_{lat} - B_{lat} K_{\mathbf{x}_{lat}}) \mathbf{x}_{lat}$$
$$+B_{lat}[\frac{K_{V_y}}{s}[(k_{Py} + k_{Dy}s)[y_{ref} - y] - V_y] - \frac{K_{\beta_{lat}}}{s}[\beta]]$$
$$+F_L L_{wake}(y, z) + F_V V_{wake}^{CL}(y, z)$$
$$y = C_y \mathbf{x}_{lat}, \tag{9.21}$$

where $K_{\beta_{lat}} = (K_{\beta_{31}} \quad K_{\beta_{41}})^T$. Now let the transfer functions in free flight (with all wake terms zero) from the reference positions to the position be:

$$y(s) = F_{i1}(s)y_{ref}(s) \tag{9.22}$$
$$z(s) = F_{i2}(s)z_{ref}(s). \tag{9.23}$$

2. Since the autopilot has been designed to be asymptotically stable at all points in the wake, we can write the following transfer function representations from the linearizations of Eqns. (9.20), and (9.21) at a point

$(\bar{y}, \bar{z})$ in the wake:

$$\delta y(s) = F_{i1}(s)(1 + \Delta_1(s))\delta y_{ref}(s) \qquad (9.24)$$
$$\delta z(s) = F_{i2}(s)(1 + \Delta_2(s))\delta z_{ref}(s), \qquad (9.25)$$

where the uncertain transfer functions $\Delta_1(s)$ and $\Delta_2(s)$ arise from the wake feedback nonlinearities.

3. Use the free flight dynamics with the autopilot to estimate the contribution to pitch angle θ_z of the vertical separation reference signal z_{ref}:

$$\dot{\hat{\mathbf{x}}}_{long} = \left(A_{long} - B_{long}K_{\mathbf{x}_{long}}\right)\hat{\mathbf{x}}_{long}$$
$$+B_{long}[\frac{K_{V_x}}{s}[(k_{Px} + k_{Dx}s)[x_{ref} - x] - V_x]]$$
$$+B_{long}[\frac{K_{V_z}}{s}[(k_{Pz} + k_{Dz}s)[z_{ref} - z] - V_z]]$$
$$\theta_z = C_\theta\hat{\mathbf{x}}_{long}, \qquad (9.26)$$

Subtract θ_z from θ to estimate the pitch angle due to the upwash θ_w:

$$\dot{\mathbf{e}}_{long} = \left(A_{long} - B_{long}K_{\mathbf{x}_{long}}\right)\mathbf{e}_{long} + F_W\bar{W}_{wake}(y, z)$$
$$\theta_w = C_\theta\mathbf{e}_{long}, \qquad (9.27)$$

where
$\mathbf{e} = \mathbf{x} - \hat{\mathbf{x}}$. If we define $F_o(s) = C_\theta\left(s\mathbf{I} - \left(A_{long} - B_{long}K_{\mathbf{x}_{long}}\right)\right)^{-1}F_W$, the linearization of Eqn. (9.27) at some point $(\bar{y}, \bar{z})$ in the wake yields $\delta\theta_w = F_o(s)(1 + \Delta_o(s))\delta z_{ref}(s)$ where $\Delta_o(s)$ depends upon the gradient of the wake field at $(\bar{y}, \bar{z})$ and $F_o(s)(1 + \Delta_o(s))$ is exponentially stable at all points in the wake from autopilot design. Finally, closed loop aircraft dynamics in $F_o(s)$ are much faster than those in the position tracking dynamics.

4. Using the fact that the wake nonlinearities are bounded, we use the following representations for the purpose of extremum seeking design: $y(s) = F_{i1}(s)(y_{ref}(s) + n_y)$, $z(s) = F_{i2}(s)(z_{ref}(s) + n_z)$ and $\theta_w(s) = F_o(s)(\bar{W}_{wake}(y, z))$. Treatment of the wake terms as bounded noise does not alter performance of the extremum seeking scheme [5]. The finite range of slopes (from 0 to a maximum value) of the wake nonlinearities, and the small variation of system poles during motion in the wake due to the high gain autopilot design ensures that the dynamics are linearly stable at all points in the wake.

5. The wake nonlinearity $\bar{W}_{wake}(y, z)$ maps the outputs of the transfer functions $F_{i1}(s)$ and $F_{i2}(s)$ to the input of $F_o(s)$. Since it has a minimum, it can be represented locally around the minimum in the form in Eqn. (1.13)

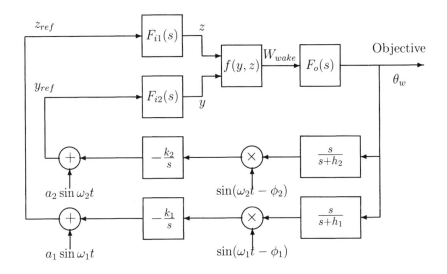

Figure 9.18: Equivalent block diagram for extremum seeking design.

with piecewise constant θ^* and f^*. Hence, we can write the closed loop adaptive system in Figure 9.17 in the form in Figure 9.18 for the purpose of extremum seeking design. An alternative means of transforming the system for design of stable extremum seeking is to directly analyze the closed loop adaptive system using the methods of averaging and singular perturbation, a lengthy procedure.

Thus, we have a system that satisfies the conditions in of Theorem 2.8 under which the design algorithm for extremum seeking can be applied. For the purpose of design, we use the transfer functions in free flight as the nominal system and perform extremum seeking design upon it in the next subsection taking into account the wake-induced uncertainty. This is justified by the fact that operation in the wake produces only small changes in the closed loop dynamics with autopilot.

9.3.2   Extremum Seeking Design for Formation Flight

We observe that the extremum seeking design for minimum power demand formation flight must, for practical implementation, satisfy the following requirements: achieve stable tracking of the optimal position from afar (at least as much as the uncertainty in vortex position) in the face of wake induced uncertainty in the transfer functions $\Delta_1(s)$, $\Delta_2(s)$ and $\Delta_o(s)$ and the map second derivative; converge to the extremum fast enough to enable maximal extraction of formation benefits under varying conditions; avoid positioning the wingman far into the downwash region of the leader (where control authority may not be sufficient to stabilize aircraft) by avoiding overshoot in the transient response, and provide reasonable robustness of performance to unexpected atmospheric turbulence.

We design two extremum seeking loops: one for attaining optimal vertical separation, and the other for attaining optimal lateral separation between the aircraft. For the process of design, we assume step variations in the optimal separations, i.e., $\Gamma_y(s) = \Gamma_z(s) = 1/s$, and in the magnitude of average upwash velocity at the optimal position $\Gamma_{\bar{W}_{wake}}(s) = 1/s$. We also simplify the design for $C_{ip}(s)$ by setting $\phi_p = -\angle(F_{ip}(j\omega_p))$, and obtaining $X_{pq}(s) = \frac{a_p P_{pq}}{2}|F_{ip}(j\omega_p)|H_{ip}(s)$.

We first apply the Design Algorithm 2.2.1 to design of the vertical separation optimization loop that sets the reference z_{ref} for the longitudinal aircraft dynamics with autopilot. We choose forcing frequency $\omega_1 = 3$ rad/s (about twice the speed of the dominant poles of the longitudinal dynamics with autopilot) to ensure separation of time scales. Forcing amplitude $a_1 = 0.1/|F_{i1}(j\omega_1)| = 1.22$ ft is chosen so as to achieve an oscillation of 0.1 ft in aircraft vertical separation z. The output compensator is chosen as $C_{o1}(s) = 1/(s+h_1)$ with $h_1 = \omega_1 = 3$ to achieve washout action. The phase of the demodulation signal is chosen as $\phi_1 = -\angle F_{i1}(j\omega_1) = -1.8$ rad. Finally, the input compensator is chosen as a simple gain $C_{i1}(s) = k_1 = 700$.

Next, we apply the Design Algorithm 2.2.1 to design of the lateral separation optimization loop that sets the reference y_{ref} for the lateral-directional aircraft dynamics with autopilot. We choose forcing frequency $\omega_2 = 1.5$ rad/s (about twice the speed of the dominant poles of the lateral-directional dynamics with autopilot) to ensure separation of time scales. Forcing amplitude $a_2 = 0.1/|F_{i2}(j\omega_2)| = 1.58$ ft is chosen so as to achieve an oscillation of 0.1 ft in aircraft lateral separation y. The output compensator is chosen as $C_{o2}(s) = 1/(s+h_2)$ with $h_1 = \omega_2 = 1.5$ to achieve washout action. The phase of the demodulation signal is chosen as $\phi_2 = -\angle F_{i2}(j\omega_2) = 1.45$ rad. Finally, the input compensator is chosen as a simple gain $C_{i2}(s) = k_2 = 175$.

9.4   Simulation Study

We present here two sets of simulation results showing time trajectories of relative position, extremum seeking objective θ_W, and actuator and engine outputs: one in calm air (Fig. 9.19), and the other showing a brief encounter with clear air turbulence, or CAT (Fig. 9.20). For the simulation in calm air, such as that for the autopilot simulation run, the wingman is initially 20 ft below and 20 ft to the right of the optimal position (this will in practice be the best available estimate of the optimal position due to the uncertainty introduced by vortex transport [49]). The simulation of the brief encounter with CAT starts in calm air (this is typical, as the system would not be

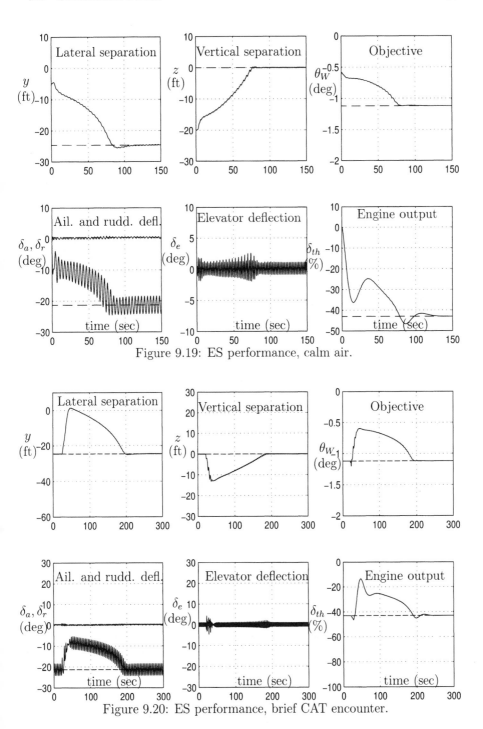

Figure 9.19: ES performance, calm air.

Figure 9.20: ES performance, brief CAT encounter.

used in known turbulent conditions), at the optimal position with turbulence beginning at 40 seconds.

The states of the longitudinal dynamics, with the exception of the elevator servo states and the engines state, are initialized close to their trim condition in the wake. This initialization, or trimming, is essential for stable functioning of the system as extremum seeking offers only a local stability guarantee. No such initialization is needed for the lateral-directional dynamics as the objective θ_w is a function of states in the longitudinal dynamics only. The turbulence model includes vertical and lateral gusts with standard Dryden spectrum for CAT at 40000 ft. The plots show steady state values in dashed lines and system performance in solid lines.

The overall result is that engines output is in the neighborhood of steady-state reduction after 80 seconds, although convergence is complete only after about 120 seconds. The speed of convergence of the adaptation is ultimately limited by the speed of the aircraft dynamics with autopilot. Here, the gains usable in extremum seeking design are limited by the presence of nonminimum phase zeros in the aircraft dynamics. A possible solution to this problem may be to use direct lift control. The simulation results show several aspects of the design that render practical application of the design procedure in this chapter feasible:

1. Ensures that there is no overshoot in the lateral separation (y) tracking. This is essential to prevent the wingman from entering the downwash region of the leader's wake where roll-control authority may not be sufficient to stably maintain aircraft position.

2. The amplitude of steady-state pitch angle oscillation θ_w is 0.2 degrees, which is sufficient for accurate measurement.

3. Steady-state actuator oscillations are reasonable: elevator oscillations are about 2 degrees at a frequency of 3 rad/s; ailerons and rudder oscillations are, respectively, about 4 degrees, and 1 degree at a frequency of 1.5 rad/s. To reduce actuator wear due to the probing signals, extremum seeking can be switched on or off using dead-zone nonlinearities before the extremum seeking integrators depending upon the distance from the optimum.

4. Actuators do not hit saturation.

The performance in clear air turbulence (CAT) reveals some fundamental limitations of the extremum seeking method. In fact, the average upwash in the wake has a peak of about 15 ft/s, and the CAT has velocity fluctuations of the order of 10 ft/s. This produces several transient local maxima in the upwash field that mislead the extremum seeking algorithm, which is based essentially on gradient estimation of the upwash field. A practical solution to this problem is switching off the extremum seeking when the vertical acceleration exceeds an acceptable level of $0.2g$. This approach was implemented with the addition of a relay and two switches to the scheme, and has been successfully tested in simulation (the results are not shown here due to space limitations).

This study did not consider trim drag due to flight in a wake. Aileron trim drag, which is the most persistent, can be balanced by asymmetric fuel loading—more fuel on the left wing tank. If the aircraft has an aft fuel tank, fuel shifting can be applied to cancel elevator trim drag; in any case this contribution is negligible, because the angle of attack change from free to formation flight is almost zero. Finally rudder trim drag can be eliminated using slightly asymmetric thrust.

Notes and References

The model-based open-loop approach for minimizing power demand in formation has been employed in prior work [45, 91, 101]. Its effectiveness is limited by the uncertainty of aerodynamic interference modeling, accompanied by high sensitivity of power demand reduction to positioning: an error of just 10% of the aircraft wingspan can reduce the benefits by half [22]. The need for accurate steady state performance in the presence of modeling uncertainty calls for adaptive feedback control. This has been done through extremum seeking algorithms [27, 56]. In [56], a simple discrete time extremum seeking algorithm to maximize aileron deflection was used to attain a power demand reduction of 20% in experimental flight tests of two Dornier aircraft in formation. In [27], simulation studies of a continuous time extremum seeking algorithm to maximize induced lift were presented. This chapter is based upon [20].

Our design can be extended to the case of maneuvering flight. The architecture of the lateral-directional part of the autopilot has to be changed, in order to track the heading reference signals, instead of lateral separation signals. An outer loop has also to be added to this and can be designed as in [91] . The longitudinal part of the autopilot does not need any modification. The extremum seeking algorithm used is capable of tracking general time variations in the location and the value of the maximum upwash velocity. Designs for formations involving more aircraft will need to consider issues of string stability.

Chapter 10

Combustion Instabilities

Lower emission requirements have motivated development of lean premixed combustors for industrial gas turbines. Their susceptibility to thermoacoustic pressure oscillations, and subsequent decreased durability have motivated a large body of research on combustion instability control. Prior experimental results and model-based analysis show that pressure measurement and a simple phase-shifting controller with an appropriately chosen phase-shift to actuate either fuel-injection or a loudspeaker is sufficient for suppression of oscillations, given enough control authority [12, 31, 52, 71]. The difficulty in determining the optimal phase shift that minimizes pressure oscillations, either by analysis or by experiment, especially in large industrial-scale combustors that operate over a wide range of conditions, has led researchers to call for the use of adaptive schemes [103].

This chapter provides an adaptive scheme to find the optimal phase shift online (from pressure measurement to fuel-injection), that is based on extremum seeking and motivated by physical modeling; identifies a closed loop model with phase-shifting control of combustion instability from experimental data; supplies stability analysis of the adaptive scheme based upon the identified model; develops stable extremum seeking designs and provides the first successful result on oscillation minimization in an industrial-scale 4MW gas turbine combustor[1]. The algorithm achieved the objective of monotonically reducing oscillation amplitudes below uncontrolled levels from all initial conditions. An Extended Kalman Filter [72] based frequency tracking observer [15] was used to reliably detect the in-phase and quadrature components of the dominant bulk mode of pressure oscillation over a wide range of operating conditions (bulk mode frequency varying from 150Hz to 250Hz). This prevented other frequencies and noise from entering the phase-shifting feedback. The control phase was updated using classical perturbation-based extremum seeking (Chapter 1) while the control gain was fixed.

---

[1]Conducted on a single nozzle rig at United Technologies Research Center (UTRC) in August 1998. Followed by experiments in [61] and [86].

The chapter is organized as follows: Section 10.1 presents experimental closed loop identification of an averaged model of pressure magnitude dynamics; Section 10.2 presents control-phase tuning by extremum seeking along with stability analysis, Section 10.3 presents adaptive oscillation attenuation on the 4MW single nozzle rig, and Section 10.4 discusses instability suppression during engine transient conditions.

10.1   Identification of Averaged Pressure Magnitude Dynamics

Closed loop identification experiments were conducted with the purpose of identifying an average model of pressure magnitude dynamics:

$$\dot{x}_0 = -\alpha(\theta_c)(x_0 - g(\theta_c)) \qquad (10.1)$$

$$A = x_0 + \nu, \qquad (10.2)$$

where $\alpha(\theta_c) > 0$ are the time constants of the exponential relaxation processes at the equilibria $A = g(\theta_c)$ of (10.1) , A is the measured instantaneous pressure magnitude, and ν is a colored noise modeling pressure magnitude fluctuation (due to turbulent velocity fluctuation in the nozzle). The form of this model was motivated both by model analysis and by the nature of experimental data. In experiments the control phase input $\theta_c(t)$ is provided and response of the pressure magnitude estimate $A(t)$ from the frequency tracking Extended Kalman Filter described in [15] is recorded. The equilibrium map $g(\theta_c)$ is obtained by fitting a curve through the equilibria found from steady state experiments, and decay rates $\alpha(\theta_c)$ at these equilibria are estimated from cumulative averaging of several step responses (the steps are in the controller parameter) to eliminate noise. A linear colored noise model (independent of θ_c) is obtained to fit the spectrum of the fluctuating component ν in experimental data[2].

Identification of Equilibrium Map.   An experimental equilibrium map is obtained by varying θ_c in a phase ramp/staircase of discrete steps from $0°$ to $360°$: $\theta_c(t) = \theta_{st}[t]$, where $[t]$ denotes the greatest integer less than time t, and $\theta_{st} = 15°$ is the discrete increment in θ_c in each step. The duration of steps is sufficiently long (1 sec) to allow for the transients in pressure magnitude to settle down. The steady-state values are estimated by averaging the magnitude data after the transient is over. From experimental data at 60% full combustor power shown in Fig. 10.1, there does seem to be a smooth variation of the steady-state magnitudes of thermoacoustic instability with θ_c along a single

---

[2]Youping Zhang fit the output noise model using combustion data and implemented the identified model in SIMULINK.

curve, and there is a definite minimum oscillation magnitude at a certain phase. A parametrization for the static map, motivated in part by analysis [11] and in part by the shape of the experimental static map itself, is:

$$g(\theta_c) = \gamma \left\{ \frac{1 + L\sin(\theta_c - \theta^* - \pi/2)}{1 + M\sin(\theta_c - \theta^* - \pi/2)} \right\}, \qquad (10.3)$$

where θ_c is the phase of the phase shifting controller, and $g(\theta_c)$ is the steady-state oscillation magnitude. Note that the map is parametrized with only four parameters. The parameters are obtained by fitting the parametrization to the experimental data by a nonlinear least squares fit. For the experiment at 60% full combustor power, the parameters obtained were $\gamma = 0.1246$, $\eta = .7659$, $\zeta = 0.6286$, $\theta^* = -0.6094$.

Identification of Decay Rates.   Experimentally, the magnitude transients are obtained by introducing a large square wave variation in the control phase. The duration of the pulses is long enough (2 sec) to allow settling of the transients of each step, and the upper and lower values of θ_c in the pulses are chosen so as to ensure an observable difference in equilibrium magnitude $g(\theta_c)$, typically 90 degrees.

To eliminate noise in the transient measurements, a cumulative averaging of the various step responses in a given square wave response is performed. We assume that the noise is zero mean, its autocorrelation function decays to zero within half pulse period T, and that the magnitude transients settle within half pulse period T. Time traces of transient responses are averaged cumulatively to obtain the Nth estimate of the magnitude time-trace:

$$\bar{A}(t) = \frac{1}{[t/T] + 1} \sum_{i=0}^{[t/T]} A(t - iT), \qquad (10.4)$$

where $A(t)$ is the measurement of magnitude as in the previous section, $T = 2$ sec is the time period of the pulses, and $[t/T]$ denotes the greatest integer less than t/T. The transients due to the upward and downward steps are averaged separately, since they represent transients at different equilibria. The corresponding final averages are shown in Figure 10.2. In the figures, the smooth curves of the exponential fits are superimposed over the rough curves from cumulative averages.

In the case where we can measure the magnitude perfectly without noise, direct integration of Eqn. (10.1) yields $\alpha(\theta_{\text{fin}})$ as

$$\alpha(\theta_{\text{fin}}) = \frac{A(t) - A(t + s)}{\int_t^{t+s} A(\sigma)d\sigma - sg(\theta_{\text{fin}})} \qquad (10.5)$$

$\forall t$ s.t. $\left[\frac{2t}{T}\right] = \left[\frac{2(t+s)}{T}\right]$, and $\forall s < T/2$, where θ_{fin} is the final phase of the phase step. However, since we do not have noise-free data, we estimate

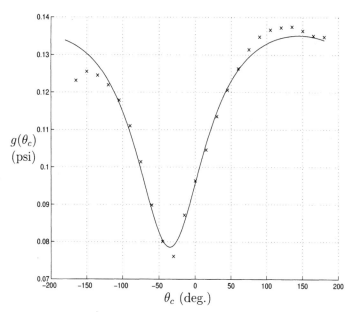

Figure 10.1: Static map from control phase θ_c to oscillation amplitude A

the exponent from the Nth cumulative average of the measured transients (Eqn. (10.4)) as follows:

$$\hat{\alpha}_N(\theta_{\text{fin}}) = \frac{\bar{A}(NT) - \bar{A}(NT + T_{\text{settle}})}{\int_{NT}^{NT+T_{\text{settle}}} \bar{A}(\sigma)d\sigma - T_{\text{settle}}g(\theta_{\text{fin}})}, \qquad (10.6)$$

Here, we approximate $\bar{A}(NT) \approx g(\theta_{\text{in}})$, and $\bar{A}(NT+T_{\text{settle}}) \approx g(\theta_{\text{fin}})$, where θ_{in} is the starting phase of the phase step, NT is a step time instant, and T_{settle} is such that the magnitude transients settle within it. The exponents $\hat{\alpha}(\theta_c)$ (θ_c in degrees) thus calculated are indicated on Figure 10.2. The identified pressure magnitude dynamics and the colored noise model have been implemented in SIMULINK for simulation studies of the adaptive algorithm to narrow the range of adjustable parameters and thereby minimize experimental time and expense.

10.2 Controller Phase Tuning via Extremum-Seeking

The optimal phase shift, being a function of the operating conditions, and being dependent upon several unknown parameters that are difficult to estimate (like heat release time delay τ) is tuned online by extremum seeking for the following reasons:

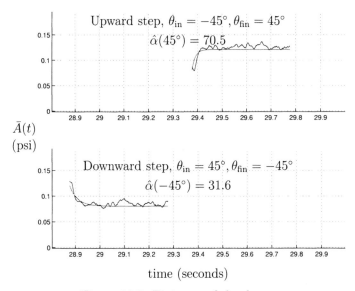

Figure 10.2: Estimate of the decay rate

1. The stable combustion process and actuator dynamics (around 200Hz) are much faster than the pressure magnitude dynamics ($\approx$ 10Hz) as verified by experiment in Section 10.1, and therefore permit the problem to be reduced to simply the reduction of the pressure amplitude (the dynamics of the frequency tracking observer are also as fast as the combustion process dynamics it observes).

2. Direct availability of the phase-shift θ_c for tuning, and the magnitude $A(t) = \sqrt{p_c(t)^2 + p_q(t)^2}$ for measurement as an objective to minimize, through the frequency tracking observer ($p_c(t)$ refers to in phase component, and $p_q(t)$ refers to the quadrature component).

3. The equilibrium map $g(\theta_c)$ of pressure amplitude versus control phase is smooth, unique, and has a unique minimum.

4. Perturbation-based extremum seeking provides guarantees of stability and convergence of θ_c to its optimum; this can be proved as in Chapter 5 using the separation of time-scales between the slow update of θ_c and the faster magnitude dynamics.

A modified version of the classical extremum seeking scheme implemented for phase-shift tuning is shown in Figure 10.3. The extremum-seeking algorithm used in this chapter relies on a small sinusoidal variation of θ_c with frequency ω and amplitude a to obtain a measure of the gradient of the map

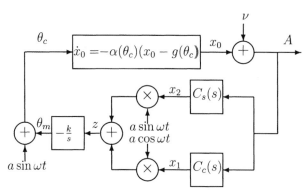

Figure 10.3: Extremum seeking scheme

$g(\theta_c)^3$. Instead of a simple washout filter, it uses a magnitude observer to extract the in-phase and quadrature components of the magnitude estimate of A at the frequency ω. The magnitude observer decomposes the magnitude estimate signal into constant, in-phase, and quadrature component (the last two relative to the phase perturbation signal $\sin(\omega t)$). The transfer functions from the magnitude estimate to the in-phase and quadrature components are given below:

$$C_s(s) = \frac{l_2 s + \omega l_1}{s^3 + (l_2 + l_3)s^2 + (\omega^2 + l_1\omega)s + l_3\omega^2} \tag{10.7}$$

$$C_c(s) = \frac{l_1 s - \omega l_2}{s^3 + (l_2 + l_3)s^2 + (\omega^2 + l_1\omega)s + l_3\omega^2}, \tag{10.8}$$

where $l_1 = -0.016$, $l_2 = 1.996$ and $l_3 = 1.996$ were chosen for stable observation. The sine and cosine components of A are demodulated by $a\sin\omega t$ and $a\cos\omega t$ respectively, and then summed up and passed through the integrator which has a gain k.

Parameter Selection. To aid selection of extremum seeking parameters ω, a and k to ensure stable extremum seeking, we performed simulation studies with extremum seeking applied to the averaged model identified in Section 10.1. Systematic design with Algorithm 1.2.1 is not possible here as the range of forcing frequencies used $(6 - 90\text{rad/s})$ overlapped with the range of plant time constants $\alpha(\theta_c) \in [15, 45]$ rad/sec in the identified models. Here, in hindsight, we can use the time-varying stability test of Chapter 1 to determine a priori the variation of performance of the extremum seeking scheme with design parameters. It can especially be used a priori to rule out destabilizing designs, or designs that are sensitive to noise frequencies in the operating environment.

---

[3]A different approach in [122] uses the triangular search algorithm, which uses three past sampled average magnitude values to determine the new control phase.

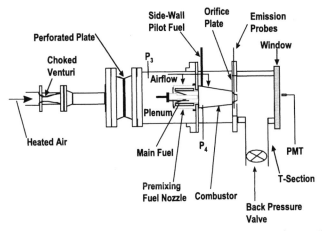

Figure 10.4: UTRC single-nozzle combustion rig (4 MW)

10.3 Experiments with the Adaptive Algorithm

A cost-effective alternative to both engine and full annular combustor testing is to test a sector cut from the full combustor annulus containing one or several fuel nozzles. In this section we present results of experiments in United Technologies Research Center conducted on 4 MW Single Nozzle Rig in August 1998 using full-scale engine fuel nozzle at realistic operating conditions. Rig schematics are presented in Figure 10.4. About 10% of the net fuel was modulated for control purposes using a linear proportional valve (For more details on the UTRC experimental rigs see [31].). The control gain is fixed and only the control phase is updated using the algorithm described in Section 10.2.

Performance specifications for the adaptive algorithm have been defined for algorithm initialization transients and engine acceleration transients: when initialized with a phase corresponding to amplification of oscillations, the algorithms should quickly produce and maintain phases corresponding to suppression of the oscillations; during engine acceleration transients the algorithms should be able to suppress oscillations relative to uncontrolled levels.

The dependence of the mean pressure magnitude and frequency of the corresponding mode on the control phase has been determined experimentally at several power conditions, so that the optimal control phase θ_c^* was known a priori. This information let us check the performance of the extremum-seeking algorithm. To test the transient performance of the adaptive algorithm, initialization transients are introduced, where the initial control phase $\theta_c(0)$ differs significantly from θ_c^*. We show time traces of control phase and pressure mag-

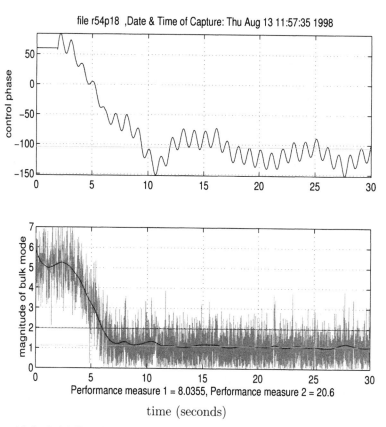

Figure 10.5: Initialization transient for higher power: Time traces of control phase (top) and pressure magnitude (bottom).

nitude as functions of time during initialization transients in Figure 10.5 for combustor operation at 73% of full power, and in Figure 10.6 for combustor operation at 60% of full power. The parameters for the extremum seeking for the higher and lower power experiments are respectively $f = 1Hz$, $a = 15°$, $k = 1000$, $\theta_m(0) = 60°$ and $f = 1Hz$, $a = 15°$, $k = 150$, $\theta_m(0) = 115°$. The pressure is in pounds per square inch. The optimal control phase is shown as a dotted line in the upper half of the figures, the bottom halves of the figures show the magnitude of the pressure oscillation amplitude with a low pass filtered average superimposed. Also shown in the bottom half of the figures are the average magnitude of oscillations without feedback control (as a solid line) and the minimal oscillation amplitude attained by extremum seeking (as a dotted line). Figures 10.7 and 10.8 show dependence of low-pass filtered pressure magnitude on the control phase overlaid over a sketch of the map representing the mean pressure magnitude dependence on the control phase from control ramp experiments. For the frequency $f = \omega/2\pi$ of sinusoidal variation

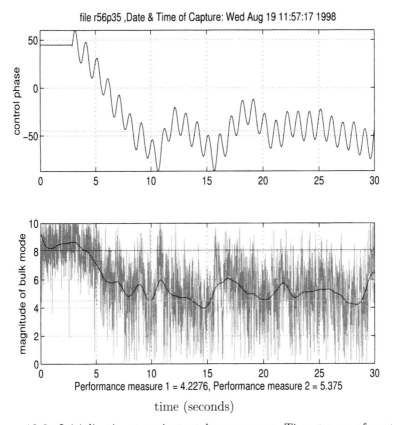

Figure 10.6: Initialization transient at lower power: Time traces of control phase (top) and pressure magnitude (bottom).

introduced in the control phase below $10Hz$ (corresponding to a separation of time-scales), integrator gain k ranging from 150 to 1000, and amplitude of forcing $a = 10°, 15°$, the algorithm behaved very well at high power condition (medium noise and small pressure oscillations) and reasonably well at low power conditions (large noise and pressure oscillations). On reaching a neighborhood of the optimal value, the control phase usually stayed in a reasonably small neighborhood of that value, rarely produced control phases corresponding to level higher than uncontrolled levels, and always provided better average pressure oscillations levels than uncontrolled levels. For further details on the experimental attenuation of pressure oscillations using the extremum seeking algorithm described in this chapter, see [14].

It has been inferred that the major factor affecting the performance of the extremum-seeking schemes is the "noise" present in the pressure magnitude. This noise component (denoted by ν) was introduced in the model in Section 10.1; the noise can be attributed to an effect of turbulent flow in the

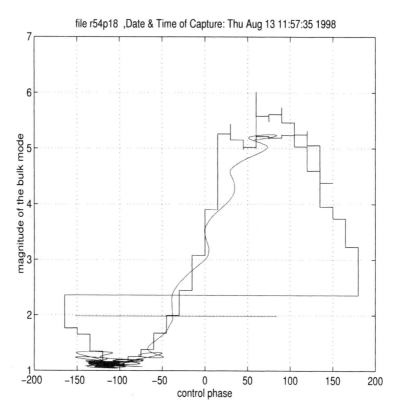

Figure 10.7: Initialization transient for higher power: pressure magnitude as a function of control phase.

nozzle. The changes in operating conditions appearing during engine acceleration and deceleration are likely to resemble the transients between different power levels on the single nozzle rig. In experiments, the frequencies of the pressure modes, the mean pressure magnitude levels, and noise levels varied significantly between power levels.

It was determined that in order to work in a simulated transient from low to high power conditions, the classical algorithm would have to be modified to allow for adaptive gain change (by a factor of five). One fixed gain k would not work at both low and high power conditions.

10.4 Instability Suppression during Engine Transient

In this section, we simulate an extremum seeking design under an engine transient upon the closed loop model identified in Section 10.1. For the purpose

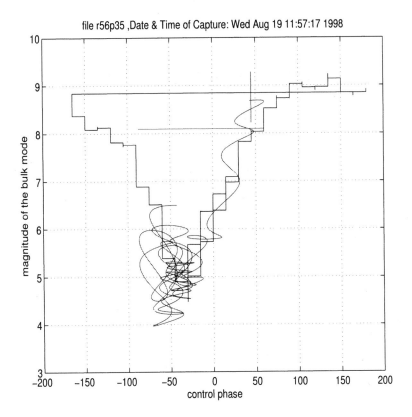

file r56p35 ,Date & Time of Capture: Wed Aug 19 11:57:17 1998

Figure 10.8: Initialization transient at lower power: pressure magnitude as a function of control phase.

of our simulation example, we use parameters identified from data at 60% power and 73% power. At 60% power, $\gamma = 0.1246$, $\eta = 0.7659$, $\theta^* = -0.6094$ and $\zeta = 0.6286$. At 73% power, $\gamma = 0.0454$, $\eta = 0.4323$, $\theta^* = -0.4001$, and $\zeta = -0.2419$. The curve fit for $\alpha(\theta)$ from [100] used in the simulations is:

$$\alpha(\theta) = a_1 \sin(\theta - \theta^* + a_2) + a_3, \qquad (10.9)$$

where $a_1 = 26.15$, $a_2 = 0.6094$, $a_3 = 44.8750$. Above, we showed successful attenuation of oscillations in experiments at fixed combustor power. A problem that remains is to attain attenuation of oscillations during *fast engine transients*, where all the map parameters change. Here, we show in simulations that fast tracking can be achieved.

We simulate the engine transient as a ramping of all the model parameters in Eqn. (10.1) from their values at 60% power to those at 73% power in 5 seconds, starting at time 10 seconds. For the purpose of designing the extremum seeking controller, we use $F_i(s) = 1$, $F_o(s) = \frac{\alpha}{s+\alpha}$,

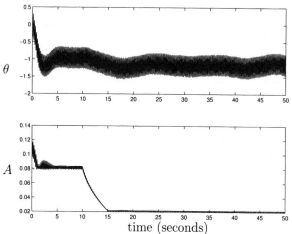

Figure 10.9: Large amplitude modulation

$\alpha \in [18, 71]$, $\Gamma_f(s) = \frac{1}{s}$, $\Gamma_\theta(s) = \frac{1}{s^2}$. We design $\omega = 30\text{rad/s}$, $C_o(s) = \frac{1}{s+3}$, and $C_i(s) = k(.01s^2 + 1.01s + 1)$ that satisfy the stability test in Proposition 1.5. We illustrate two designs, one with $k = 300$, $a = 0.3$, in Figure 10.9, and the other with $k = 600$, $a = 0.04$ in Figure 10.10 (Note that $A^*(t)$ and $\theta^*(t)$ are in dotted lines and A and θ are in solid lines). Comparison of the two simulations shows that a large amplitude modulation signal a greatly reduces noise sensitivity and produces a more oscillatory output. But as we see from the Figure 10.9, this can be acceptable since the sinusoidal forcing is greatly attenuated as it passes through the plant. Another point to be noted is that the designs are robust to the change of second derivative of the map from one operating point to another (the change in f'' here is about a factor of 10). In both cases, the algorithms were initialized at worst case amplitude $A_0 = .13$.

Notes and References

Control of thermoacoustic instabilities has of late been a focus of intense research. Many researchers have developed successful experimental controllers (see [2, 3] for references). Several control strategies have been applied to suppress thermoacoustic instabilities, some model-based, and some empirical. All the control strategies sense the pressure fluctuation in the combustor, and feed it back through actuation by loudspeaker, by an auxiliary fuel source, or by modulating the main fuel supply. Most of them have been applied to the case where the acoustics are dominated by a single harmonic, and the control objective is then to reduce its amplitude. Many of the controls, that actuate the fuel flow with time-delayed/phase-shifted measured pressure fluctuation have been motivated by the Rayleigh criterion. Extremum seeking was applied in [11]

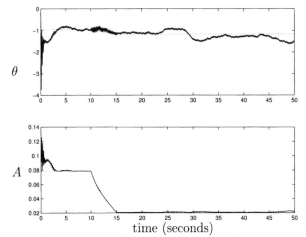

Figure 10.10: Small amplitude modulation

for the suppression of thermoacoustic instabilities in a gas turbine combustor. Extremum seeking was used to tune the phase-shift θ of a Rayleigh-criterion motivated phase-shifting controller that uses pressure measurement and actuates fuel-flow. This chapter is based on [5, 11]. The material presented can also be viewed as an application of the scheme of Chapter 6 as actuator limitations preclude complete suppression of the combustion oscillations.

Chapter 11

Compressor Instabilities: Part I

In recent years, aeroengine compressor systems have become a subject of major interest to control engineers. There are two types of instability in compressors—*rotating stall* and *surge*. While surge can lead to engine damage, rotating stall can cause a sudden drop in performance. Feedback control is necessary to avoid these two instabilities.

A basic understanding of the effects of rotating stall can be gained by considering the operating characteristic in Figure 11.1. Since, in the operation of a compressor, the objective is to increase the pressure rise Ψ, the operating point is moved along the axisymmetric characteristic to lower values of flow Φ. If the operating point is moved beyond the peak, the compressor drops into a regime with drastically reduced pressure rise. Moreover, an attempt to return immediately to the regime of high pressure is defeated by the presence of a hysteresis.

Frequencies of higher-order modes of fluid dynamic phenomena participating in rotating stall and surge far exceed the bandwidth of available actuators. For this reason, the most meaningful approach to control design is via low-order models. The simplest model that adequately describes the basic dynamics of rotating stall and surge and their interaction is the three-state nonlinear model of Moore and Greitzer [84]—MG3—which is a Galerkin approximation of a higher-order partial differential equation (PDE) model. Even though this model represents a simplification of the dynamics of a real compressor, it has been the cornerstone of some of the most successful feedback control designs which have been validated *experimentally*.

The basic MG3 model, besides having three states, models the compressor nonlinearity with a third order (cubic) polynomial. Many experimental compressors [77, 19] have non-cubic characteristics that create a deeper hysteresis in the rotating stall diagram. Figure 11.1 gives an example of an experimental characteristic [19] for a compressor in the laboratory of Richard Murray at California Institute of Technology. The deep hysteresis denoted by diamonds is not seen on a corresponding MG3 model.

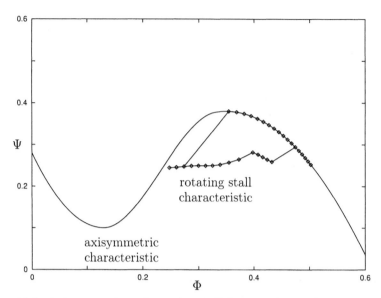

Figure 11.1: Axisymmetric and rotating stall characteristics of an experimental compressor at Caltech. The stall characteristic exhibits deep hysteresis.

It was first observed by Janković [60] that the deeper hysteresis introduces a fundamental obstacle to control design. The source of this obstacle is a (nonlinear) nonminimum-phase property not present in the MG3 model. Sepulchre and Kokotović [102] developed a "two-sine" model that can describe deep-hysteresis compressors in the region of nominal operation and explored conditions for feedback stabilization of these compressors. However, [102] uses Bessel functions, which prevents easy comparison with the basic MG3 model and its bifurcations [80].

In this chapter we present a new parameterization which is based on a convex combination of the cubic compressor characteristic and another simple but non-polynomial function. With this combination, we achieve deeper hysteresis compressor characteristics which can be used to model compressors with the deep-hysteresis property. The model uses a single parameter ϵ to describe an entire family of compressors. We develop a family of controllers which are applicable not only to the particular ϵ-MG3 model, but also to general Moore-Greitzer type one-mode models with *arbitrary compressor characteristics*. We show that each of our controllers achieves a supercritical (soft) bifurcation, that is, instead of an abrupt drop into rotating stall, it guarantees a gentle descent with a small stall amplitude.

Deep-hysteresis compressors were dubbed *right-skew* by Art Krener, and low-hysteresis compressors were termed *left-skew*. We shall use this terminology here. The stabilization designs we present in this chapter require minimal

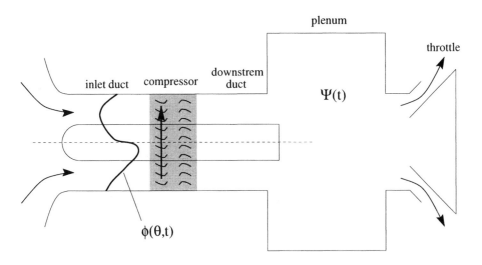

Figure 11.2: Compression system.

modeling knowledge—only the angles of arrival of the stall characteristic at the stall inception point—which are easy to determine from experimental bifurcation diagrams. Our results confirm experimentally observed difficulties for control of compressors that have a high value of Greitzer's B parameter.

As the main result of this chapter, we employ extremum seeking to achieve maximization of the compressor pressure rise in the presence of an uncertain "compressor characteristic" whose argument of the maximum is unknown. Actuation (and sensing) requirements for extremum seeking are much less demanding than those for stabilization of stalled equilibria.

This chapter is organized as follows. In Section 11.1 we derive a model that we refer to as the ϵ–MG3 model and in Section 11.2 we show that its equilibrium structure exhibits a deep hysteresis. Section 11.3 contains a formal definition of the notion of skewness and a derivation of the skewness as a function of ϵ. In Section 11.4 we study stability of equilibria by computing bifurcation diagrams. In Section 11.5 we prepare the terrain for the control design by defining the so-called "critical slopes." Section 11.6 presents and analyzes control designs which stabilize stalled equilibria. Finally, Sections 11.7, 11.8, and 11.9 present extremum seeking control to tune a stabilizing controller for maximum pressure rise. We relegate some details of technical derivations to the appendices.

Table 11.1: Notation in the Moore-Greitzer model

$\Phi = \hat{\Phi}/W - 1 - \Phi_{C0}$	$\hat{\Phi}$ — annulus-averaged flow coefficient W — compressor characteristic semi-width
$\Psi = \hat{\Psi}/H$	$\hat{\Psi}$ — plenum pressure rise H — compressor characteristic semi-height
$A = \hat{A}/W$	$\hat{A}$ — rotating stall amplitude
Φ_T	mass flow through the throttle$/W - 1$
θ	angular (circumferential) position
$\beta = \dfrac{2H}{W}B$	B — Greitzer stability parameter
$\sigma = \dfrac{3l_c}{m + \mu}$	l_c — effective length of inlet duct normalized by compressor radius m — Moore expansion parameter μ — compressor inertia within blade passage
$t = \dfrac{H}{W l_c}\hat{t}$	$\hat{t}$ — (actual time) × (rotor angular velocity)

11.1 Model Derivation

The simplest model that adequately describes the dynamics of rotating stall and surge in axial-flow compression systems shown in Figure 11.2 is the three-state Moore-Greitzer model [84]:

$$\dot{A} = \frac{\sigma}{3\pi}\int_0^{2\pi}\Psi_C\left(\Phi + A\sin\theta\right)\sin\theta d\theta \qquad (11.1)$$

$$\dot{\Phi} = -\Psi + \frac{1}{2\pi}\int_0^{2\pi}\Psi_C\left(\Phi + A\sin\theta\right)d\theta \qquad (11.2)$$

$$\dot{\Psi} = \frac{1}{\beta^2}\left(\Phi - \Phi_T\right). \qquad (11.3)$$

The quantities appearing in this model are listed in Table 11.1.

Equation (11.3) is mass conservation in the plenum: the time rate of change in plenum pressure is proportional to the difference between mass flow entering and exiting the plenum. Equation (11.2) is a momentum balance: the acceleration of the fluid in the upstream and downstream ducts is proportional to the difference in the pressure rise across the compressor and the pressure rise in the plenum. The steady-state, annulus-averaged pressure rise across the compressor is given via the S-shaped compressor characteristic $\Psi_C(\cdot)$ shown as the solid curve in Figure 11.1. Equation (11.1) is obtained from the same momentum balance PDE as (11.2) by applying the Galerkin approximation.

A standard compressor characteristic introduced by Moore and Greitzer is the cubic characteristic

$$\Psi_C(\Phi) = \Psi_{C0} + 1 + \frac{3}{2}\Phi - \frac{1}{2}\Phi^3. \tag{11.4}$$

This characteristic is adequate only for left-skew compressors. We replace it by a convex combination of a cubic characteristic (11.4) and the function $\frac{2\Phi}{1+\Phi^2}$:

$$\boxed{\Psi_C(\Phi) = \Psi_{C0} + 1 + (1-\epsilon)\left(\frac{3}{2}\Phi - \frac{1}{2}\Phi^3\right) + \epsilon\frac{2\Phi}{1+\Phi^2}} \tag{11.5}$$

where $\epsilon \in [0,1]$. The function $\frac{2\Phi}{1+\Phi^2}$ is carefully chosen so that:

1. the integrals in (11.1) and (11.2) have a closed-form solution,

2. it exhibits the qualitative properties of right-skew compressors,

3. it retains a connection with the familiar cubic characteristic. In particular, both $\frac{3}{2}\Phi - \frac{1}{2}\Phi^3$ and $\frac{2\Phi}{1+\Phi^2}$ achieve extrema ± 1 at $\Phi = \pm 1$.

Substituting $(\Phi + A\sin\theta)$ as the argument of Ψ_C, we get

$$\begin{aligned}
&\Psi_C(\Phi + A\sin\theta)\\
=\ &\Psi_{C0} + 1 + (1-\epsilon)\left(\frac{3}{2}\Phi - \frac{1}{2}\Phi^3 + \frac{3}{2}A\left(1-\Phi^2\right)\sin\theta\right.\\
&\left. -\frac{3}{2}A^2\Phi\sin^2\theta - \frac{1}{2}A^3\sin^3\theta\right) + \epsilon\frac{2(\Phi + A\sin\theta)}{1+(\Phi+A\sin\theta)^2} \tag{11.6}
\end{aligned}$$

which we use to compute the integrals in (11.1) and (11.2). First, the integral in (11.2) reduces to

$$\begin{aligned}
&\int_0^{2\pi} \Psi_C(\Phi + A\sin\theta)\,d\theta\\
=\ &2\pi(\Psi_{C0} + 1) + 2\pi(1-\epsilon)\left(\frac{3}{2}\Phi - \frac{1}{2}\Phi^3\right)\\
&-\frac{3}{2}\pi A^2\Phi + \epsilon\int_0^{2\pi}\frac{2(\Phi + A\sin\theta)}{1+(\Phi+A\sin\theta)^2}\,d\theta. \tag{11.7}
\end{aligned}$$

Using MATHEMATICA (see Appendix D for details), we get

$$\begin{aligned}
&\int_0^{2\pi} \Psi_C(\Phi + A\sin\theta)\,d\theta\\
=\ &\frac{2\sqrt{2}\pi\,\mathrm{sgn}(\Phi)}{\left[(\Phi^2 - A^2 - 1)^2 + 4\Phi^2\right]^{1/2}}\\
&\times\left\{\left[(\Phi^2 - A^2 - 1)^2 + 4\Phi^2\right]^{1/2} + (\Phi^2 - A^2 - 1)\right\}^{1/2}. \tag{11.8}
\end{aligned}$$

Second, the integral in (11.1) reduces to

$$\int_0^{2\pi} \Psi_{\text{C}} \left(\Phi + A\sin\theta\right) \sin\theta d\theta$$

$$= \; (1-\epsilon)\left(\frac{3}{2}\pi A(1-\Phi^2) - \frac{3}{8}\pi A^3\right) + \epsilon \int_0^{2\pi} \frac{2(\Phi + A\sin\theta)\sin\theta}{1 + (\Phi + A\sin\theta)^2}\,d\theta \tag{11.9}$$

With MATHEMATICA (see Appendix D), we get

$$\int_0^{2\pi} \Psi_{\text{C}} \left(\Phi + A\sin\theta\right) \sin\theta d\theta$$

$$= \; \frac{2\pi}{A}\left\{ 2 - \frac{\sqrt{2}}{\left[(\Phi^2 - A^2 - 1)^2 + 4\Phi^2\right]^{1/2}}\right.$$

$$\times \left[\left(\left(\left(\Phi^2 - 1\right)\left(\Phi^2 - A^2 - 1\right) + 4\Phi^2\right)^2 + 4\Phi^2 A^4\right)^{1/2}\right.$$

$$\left.\left. + \left(\Phi^2 - 1\right)\left(\Phi^2 - A^2 - 1\right) + 4\Phi^2\right]^{1/2}\right\}. \tag{11.10}$$

Defining

$$R = \left(\frac{A}{2}\right)^2 \tag{11.11}$$

and substituting (11.7)–(11.10) into (11.1)–(11.3), the compressor model becomes

$$\dot{R} = \sigma\left\{(1-\epsilon)R\left(1 - \Phi^2 - R\right)\right.$$

$$+ \frac{2\epsilon}{3}\left[1 - \frac{1}{\sqrt{2}\left[(\Phi^2 - 4R - 1)^2 + 4\Phi^2\right]^{1/2}}\right.$$

$$\times \left(\left(\left(\left(\Phi^2 - 1\right)\left(\Phi^2 - 4R - 1\right) + 4\Phi^2\right)^2 + 64\Phi^2 R^2\right)^{1/2}\right.$$

$$\left.\left.\left. + \left(\Phi^2 - 1\right)\left(\Phi^2 - 4R - 1\right) + 4\Phi^2\right)^{1/2}\right]\right\} \tag{11.12}$$

$$\dot{\Phi} = -\Psi + \Psi_{\text{C0}} + 1 + (1-\epsilon)\left(\frac{3}{2}\Phi - \frac{1}{2}\Phi^3 - 3\Phi R\right)$$

$$+ \epsilon\frac{\sqrt{2}\text{sgn}(\Phi)}{\left[(\Phi^2 - 4R - 1)^2 + 4\Phi^2\right]^{1/2}}$$

$$\times \left\{\left[\left(\Phi^2 - 4R - 1\right)^2 + 4\Phi^2\right]^{1/2} + \left(\Phi^2 - 4R - 1\right)\right\}^{1/2} \tag{11.13}$$

$$\dot{\Psi} = \frac{1}{\beta^2}\left(\Phi - \Phi_{\text{T}}\right). \tag{11.14}$$

We refer to this model as the ϵ–MG3 model. Note that, even though (11.13) contains sgn (Φ), this equation is not discontinuous because the term multiplied by sgn(Φ) vanishes at $\Phi = 0$ for all values of R. For $\epsilon = 0$, the model (11.12)–(11.14) reduces to the standard MG3 model

$$\dot{R} = \sigma R \left(1 - \Phi^2 - R \right) \tag{11.15}$$

$$\dot{\Phi} = -\Psi + \Psi_{C0} + 1 + \frac{3}{2}\Phi - \frac{1}{2}\Phi^3 - 3\Phi R \tag{11.16}$$

$$\dot{\Psi} = \frac{1}{\beta^2}\left(\Phi - \Phi_T \right), \tag{11.17}$$

where the quantities are defined in Table 11.1, and the throttle flow Φ_T is related to the pressure rise Ψ through the throttle characteristic

$$\Psi = \frac{1}{\gamma^2}\left(1 + \Phi_{C0} + \Phi_T \right)^2 . \tag{11.18}$$

The parameter ϵ spans an entire family of compressor models whose equilibrium structure we now study.

11.2 Equilibria in the ϵ-MG3 Model

There are two sets of equilibria of the model (11.12)–(11.14). The no-stall equilibria are:

$$\begin{bmatrix} R \\ \Phi \\ \Psi \end{bmatrix}_e = \begin{bmatrix} 0 \\ \Phi_0 \\ \Psi_C(\Phi_0) \end{bmatrix}, \qquad \Phi_0 \in \mathbb{R}. \tag{11.19}$$

The stall equilibria are:

$$\begin{bmatrix} R \\ \Phi \\ \Psi \end{bmatrix}_e = \begin{bmatrix} R_0 \\ \Phi_{R\pm}(R_0) \\ \Psi_{R\pm}(R_0) \end{bmatrix}, \qquad R_0 \in [0, \bar{R}]. \tag{11.20}$$

The functions $\Phi_{R+}(R)$ and $\Phi_{R-}(R)$ are obtained as solutions of (11.12) with $\dot{R} = 0$ and $R \neq 0$,

$$0 = (1-\epsilon)\left(1 - \Phi^2 - R \right) + \frac{2\epsilon}{3R}\Bigg\{ 1 - \frac{1}{\sqrt{2}\left[(\Phi^2 - 4R - 1)^2 + 4\Phi^2 \right]^{1/2}}$$

$$\times \left[\left(\left((\Phi^2 - 1)(\Phi^2 - 4R - 1) + 4\Phi^2 \right)^2 + 64\Phi^2 R^2 \right)^{1/2} \right.$$

$$\left. + (\Phi^2 - 1)(\Phi^2 - 4R - 1) + 4\Phi^2 \right]^{1/2} \Bigg\} . \tag{11.21}$$

Note that since the expression in (11.21) is a function of Φ^2, we get two solutions $\Phi_{R-}(R) = -\Phi_{R+}(R)$. The functions $\Psi_{R+}(R)$ and $\Psi_{R-}(R)$ are obtained as solutions of (11.13) with $\dot{\Phi} = 0$, that is, by substituting $\Phi = \Phi_{R\pm}(R)$ into

$$
\begin{aligned}
\Psi &= \Psi_{C0} + 1 + (1 - \epsilon)\left(\frac{3}{2}\Phi - \frac{1}{2}\Phi^3 - 3\Phi R\right) \\
&\quad + \epsilon \frac{\sqrt{2}\,\mathrm{sgn}\,(\Phi)}{\left[(\Phi^2 - 4R - 1)^2 + 4\Phi^2\right]^{1/2}} \\
&\quad \times \left\{\left[(\Phi^2 - 4R - 1)^2 + 4\Phi^2\right]^{1/2} + (\Phi^2 - 4R - 1)\right\}^{1/2}. \quad (11.22)
\end{aligned}
$$

The plots of equilibria of the ϵ–MG3 model are given in Fig. 11.3 ($\epsilon = 0$), Fig. 11.4 ($\epsilon = 0.5$), Fig. 11.5 ($\epsilon = 0.8$), and Fig. 11.6 ($\epsilon = 0.9$). The corresponding $R(\Phi)$ curves and $R(\Psi)$ curves are shown in Fig. 11.7 and Fig. 11.8, respectively.

The quantity $\bar{R}$ in (11.20) represents the maximum equilibrium value of R. As we can see from Fig. 11.7, $\bar{R}$ increases as ϵ increases. Furthermore,

$$
\lim_{\epsilon \to 0} \bar{R}(\epsilon) = 1 \tag{11.23}
$$

$$
\lim_{\epsilon \to 1} \bar{R}(\epsilon) = \infty. \tag{11.24}
$$

An important question is how to determine the value of ϵ from steady-state experimental data for a given compressor. Our approach is to select ϵ to match the axisymmetric characteristic. For example, $\epsilon = 0.83$ provides a reasonable match of the axisymmetric characteristic in Figure 11.1. However selecting a value of ϵ to match the shape of the axisymmetric characteristic does not guarantee that the stall characteristic will be perfectly matched. The resulting difference between the ϵ-MG3 model and an actual compressor would be primarily due to *neglected higher-order modes* of rotating stall. The contribution of higher-order modes is to further deepen the hysteresis [1, 77]. Despite the imperfect match of the stall characteristic with the one-mode ϵ-MG3 model, this is the best match possible for a three-state Moore-Greitzer type model with a sinusoidal mode of rotating stall.

11.3   Skewness

A key property affecting the ability to design feedback controllers for compressor models is the "skewness" of the compressor characteristics. To formally define the notion of skewness, we consider the stall diagrams in Fig. 11.7. We define the skewness of a stall characteristic as

$$
\mathcal{S} = \left.\frac{d\Phi_{R+}(R)}{dR}\right|_{R=0}. \tag{11.25}
$$

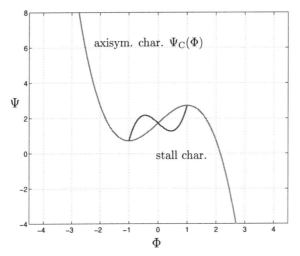

Figure 11.3: Equilibria of the ϵ–MG3 model with $\epsilon = 0$.

A compressor whose diagram at $(R = 0,\ \Phi = 1)$ has a negative slope is said to be "left-skew", while a compressor with a positive slope is referred to as "right-skew."

Let the relation between R and Φ be given by the implicit expression $\mathcal{F}(R, \Phi) = 0$ in (11.21). Then it is easy to show that

$$
\mathcal{S} = -\left.\frac{\dfrac{\partial \mathcal{F}(R,\Phi)}{\partial R}}{\dfrac{\partial \mathcal{F}(R,\Phi)}{\partial \Phi}}\right|_{\substack{R=0 \\ \Phi=1}}. \tag{11.26}
$$

Substituting (E.7) and (E.8) from Appendix E, we get

$$
\mathcal{S} = -1.5\frac{(\epsilon - 0.5)}{\epsilon - 1.5}. \tag{11.27}
$$

Thus a compressor is right skew when $\epsilon > 0.5$ and left skew when $\epsilon < 0.5$.

11.4   Open-Loop Bifurcation Diagrams

We now study stability properties of the open-loop compressor model by computing its bifurcation diagrams. Figures 11.3–11.8 show equilibria of the system obtained by setting $\dot{R} = \dot{\Phi} = 0$. By setting, in addition, $\dot{\Psi} = 0$ in (11.14), we get $\Phi = \Phi_{\rm T}$, where the flow through the throttle $\Phi_{\rm T}$ is related to the compressor pressure rise via the throttle characteristic whose form is usually

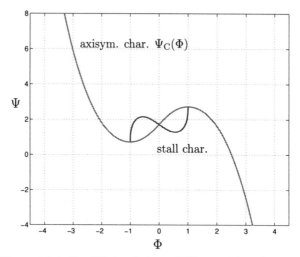

Figure 11.4: Equilibria of the ϵ–MG3 model with $\epsilon = 0.5$.

assumed to be quadratic:

$$\Psi = \frac{1}{\gamma^2}\left(1 + \Phi_{C0} + \Phi_T\right)^2, \tag{11.28}$$

where γ is the throttle opening. While Figures 11.3–11.8 show *possible* equilibria of the system, the *actual* equilibria are determined by the value of γ, in the intersection of the throttle characteristic with the compressor and stall characteristics. We now illustrate the results of the previous sections via bifurcation diagrams with γ as the bifurcation parameter.

We consider a three-stage compressor studied in [88] whose parameters are $\Psi_{C0} = 0.72$ and $\sigma = 4$. We study both a low-speed case $\beta = 0.71$ and a high-speed case $\beta = 1.42$, and both a left-skew case $\epsilon = 0$ and a right-skew case $\epsilon = 0.9$. Figure 11.9 shows a bifurcation diagram for $\epsilon = 0$, $\beta = 0.71$. The solid thick line represents stable equilibria, while the dashed thin line represents unstable equilibria. Figure 11.10 is for the right-skew case $\epsilon = 0.9$ with the same low $\beta = 0.71$. The right-skew case results in a much deeper hysteresis; while for $\epsilon = 0$ the interval of γ participating in the hysteresis is 0.17, for $\epsilon = 0.9$ it is 0.66, which indicates that a recovery from rotating stall for $\epsilon = 0.9$ would be much more difficult. There is only a slight difference between the bifurcation diagram in Figure 11.10 for $\beta = 0.71$ and that for $\beta = 1.42$. However, the *transient* behavior for the low-speed ($\beta = 0.71$) and high-speed ($\beta = 1.6$) case is quite different, with the former resulting in rotating stall while the latter results in surge, as shown in Figure 11.12. Note that the trajectory for $\beta = 0.71$ is strongly influenced by the presence of a saddle point on the axisymmetric characteristic, just left from the peak.

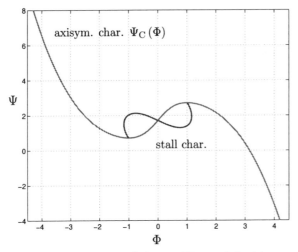

Figure 11.5: Equilibria of the ϵ–MG3 model with $\epsilon = 0.8$.

11.5   Pre-Control Analysis: Critical Slopes

A critical parameter for the control of a compressor model is the "direction" of the stall characteristic at the stall inception point. We now determine the slope of the projection of the stall characteristic to each of the three coordinate planes:

$$S_1 = \left. \frac{d\Phi_{R+}(R)}{dR} \right|_{R=0} \tag{11.29}$$

$$S_2 = \left. \frac{d\Psi_{R+}(R)}{dR} \right|_{R=0} \tag{11.30}$$

$$S_3 = \left. \frac{d\Psi_S(\Phi)}{d\Phi} \right|_{\Phi=1} = \frac{S_2}{S_1}, \tag{11.31}$$

where $\Psi_S(\Phi)$ is the stall characteristic shown in Figs. 11.3–11.6.

First, we note that

$$S_1 = \mathcal{S}. \tag{11.32}$$

Second, defining (11.22) as $\Psi = \mathcal{G}(R, \Phi)$, we get

$$
\begin{aligned}
S_2 &= \left. \frac{d\mathcal{G}(R, \Phi_{R+}(R))}{dR} \right|_{R=0} \\
&= \left. \frac{\partial \mathcal{G}(R, \Phi)}{\partial R} \right|_{\substack{R=0 \\ \Phi=1}} + \left. \frac{\partial \mathcal{G}(R, \Phi)}{\partial \Phi} \right|_{\substack{R=0 \\ \Phi=1}} S_1.
\end{aligned}
\tag{11.33}
$$

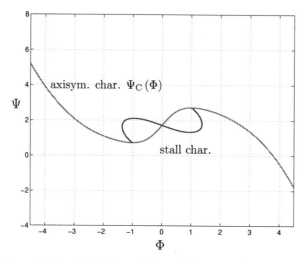

Figure 11.6: Equilibria of the ϵ–MG3 model with $\epsilon = 0.9$.

Table 11.2: Critical slopes as functions of ϵ.

		$\epsilon = 0$	$\epsilon = 1$
S_1	$-1.5\dfrac{\epsilon - 0.5}{\epsilon - 1.5}$	-0.5	1.5
S_2	$2(\epsilon - 1.5)$	-3	-1
S_3	$-\dfrac{4}{3}\dfrac{(\epsilon - 1.5)^2}{(\epsilon - 0.5)}$	6	-0.67

From (E.10), we can easily see that

$$\left.\frac{\partial \mathcal{G}(R, \Phi)}{\partial \Phi}\right|_{\substack{R=0 \\ \Phi=1}} = 0 \tag{11.34}$$

and hence,

$$S_2 = \left.\frac{\partial \mathcal{G}(R, \Phi)}{\partial R}\right|_{\substack{R=0 \\ \Phi=1}}. \tag{11.35}$$

With (E.9) we get

$$S_2 = 2(\epsilon - 1.5). \tag{11.36}$$

Table 11.2 summarizes the critical slopes as functions of ϵ.

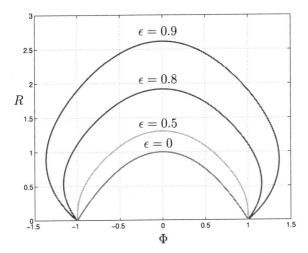

Figure 11.7: R vs. Φ relationship with varying ϵ.

11.6 Control Design

We address a general class of Moore-Greitzer type compressor models:

$$\dot{R} = \sigma R \mathcal{F}(R, \Phi) \tag{11.37}$$

$$\dot{\Phi} = -\Psi + \mathcal{G}(R, \Phi) \tag{11.38}$$

$$\dot{\Psi} = \frac{1}{\beta^2}(\Phi - \Phi_T), \tag{11.39}$$

where the functions $\mathcal{F}(R, \Phi)$ and $\mathcal{G}(R, \Phi)$ are given by

$$\mathcal{F}(R, \Phi) = \frac{1}{3\pi\sqrt{R}} \int_0^{2\pi} \Psi_C\left(\Phi + 2\sqrt{R}\sin\theta\right)\sin\theta d\theta \tag{11.40}$$

$$\mathcal{G}(R, \Phi) = \frac{1}{2\pi} \int_0^{2\pi} \Psi_C\left(\Phi + 2\sqrt{R}\sin\theta\right) d\theta. \tag{11.41}$$

The throttle flow Φ_T is related to the pressure rise Ψ through the throttle characteristic

$$\Psi = \frac{1}{\gamma^2}(1 + \Phi_{C0} + \Phi_T)^2, \tag{11.42}$$

where γ is the throttle opening. We apply control action by varying the opening γ.

A full-state feedback controller for the model (11.37)–(11.39) would employ the measurements of all three states, R, Φ, and Ψ, for feedback. In addition, the experiments in [38] show that $\dot{\Phi}$ can be measured successfully and used for feedback.

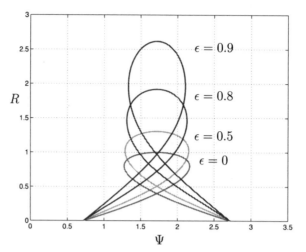

Figure 11.8: R vs. Ψ relationship with varying ϵ.

However, we are motivated to look for partial-state feedback controllers to reduce the sensing requirements. For example, for *left-skew* compressors, it was shown in [68] that stabilization is possible using a controller of the form

$$\gamma = \frac{\Gamma + \bar{\beta}^2 \left(c_\Psi \Psi - c_\Phi \Phi \right)}{\sqrt{\Psi}}, \tag{11.43}$$

i.e., without using R and $\dot{\Phi}$. As we shall see later in this chapter, controlling a right-skew compressor will require a measurement of either R or $\dot{\Phi}$. Thus, we postulate that the controller will be of the form

$$\boxed{\gamma = \frac{\Gamma + \bar{\beta}^2 \left(c_\Psi \Psi - c_\Phi \Phi + c_R R - d_\Phi \dot{\Phi} \right)}{\sqrt{\Psi}}.} \tag{11.44}$$

The controller development in this section is independent of the form of compressor characteristic. We only require that $\Psi_C'(1) = 0$ and $\Psi_C''(1) < 0$, i.e., that $\Psi_C(\Phi)$ has a maximum at $\Phi = 1$.

11.6.1   Enforcing a Supercritical Bifurcation

With the controller (11.44), the system (11.37)–(11.39) becomes

$$\dot{R} = \sigma R \mathcal{F}(R, \Phi) \tag{11.45}$$

$$\dot{\Phi} = -\Psi + \mathcal{G}(R, \Phi) \tag{11.46}$$

$$\dot{\Psi} = \frac{\bar{\beta}^2}{\beta^2} \left(-c_R R + c_* \Phi - c_\Psi \Psi + d_\Phi \dot{\Phi} \right) + \frac{1 + \Phi_{C0} - \Gamma}{\beta^2}, \tag{11.47}$$

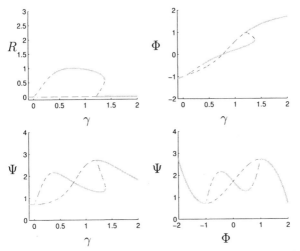

Figure 11.9: Bifurcation diagrams for the open-loop system with $\epsilon = 0$ and $\beta = 0.71$. The throttle opening γ is the bifurcation parameter.

where

$$c_* = c_\Phi + \frac{1}{\bar{\beta}^2}.\tag{11.48}$$

Let us consider an equilibrium on the stall characteristic. The equilibrium is determined by the value of Γ, which is given as

$$\Gamma(R_0) = 1 + \Phi_{C0} + \bar{\beta}^2\left[-c_R R_0 + c_*\Phi_{R+}(R_0) - c_\Psi\Psi_{R+}(R_0)\right]\tag{11.49}$$

at an equilibrium with $R = R_0$. For the bifurcation at the stall inception point to have a supercritical character with respect to Γ, we need to achieve

$$\lim_{R\to 0^+}\frac{d\Gamma(R)}{dR} < 0.\tag{11.50}$$

Noting that

$$\lim_{R\to 0^+}\frac{d\Gamma(R)}{dR} = -\bar{\beta}^2\left(c_R - S_1 c_* + S_2 c_\Psi\right),\tag{11.51}$$

we conclude that the bifurcation will be supercritical if and only if

$$c_R - S_1 c_* + S_2 c_\Psi > 0.\tag{11.52}$$

Since large Γ means lower stall amplitude R, we also require that in *no-stall* operation Φ increases with Γ (accompanied by a decreasing pressure rise Ψ). In other words, we consider an axisymmetric equilibrium with

$$\Gamma(\Phi_0) = 1 + \Phi_{C0} + \bar{\beta}^2\left(c_*\Phi_0 - c_\Psi\Psi_C(\Phi_0)\right)\tag{11.53}$$

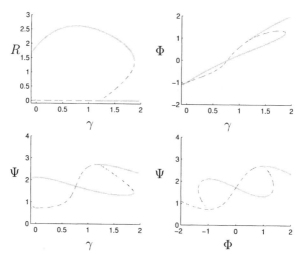

Figure 11.10: Bifurcation diagrams for the open-loop system with $\epsilon = 0.9$ and $\beta = 0.71$.

and require that

$$\lim_{\Phi \to 1^+} \frac{d\Gamma(\Phi)}{d\Phi} > 0 \,. \tag{11.54}$$

Noting that

$$\lim_{\Phi \to 1^+} \frac{d\Gamma(\Phi)}{d\Phi} = \bar{\beta}^2 c_* \,, \tag{11.55}$$

we conclude with the requirement

$$c_* > 0 \,. \tag{11.56}$$

11.6.2 Linearization at a Stall Equilibrium

We consider an equilibrium $R = R_0$, $\Phi = \Phi_{R+}(R_0)$, $\Psi = \Psi_{R+}(R_0)$ and define the error coordinates

$$r = R - R_0 \tag{11.57}$$
$$\phi = \Phi - \Phi_{R+}(R_0) \tag{11.58}$$
$$\psi = \Psi - \Psi_{R+}(R_0) \,. \tag{11.59}$$

The linearization of the system (11.45)–(11.47) is readily shown to be

$$\dot{r} = -a_1(R_0)(-\Sigma_1(R_0)r + \phi) \tag{11.60}$$
$$\dot{\phi} = (\Sigma_2(R_0) + a_2(R_0)\Sigma_1(R_0))r - a_2(R_0)\phi - \psi \tag{11.61}$$
$$\dot{\psi} = \kappa\left(-c_R r + c_*\phi - c_\Psi\psi + d_\Phi\dot{\phi}\right), \tag{11.62}$$

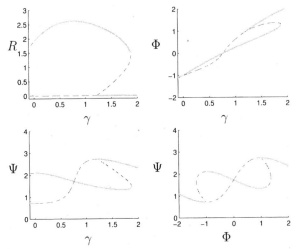

Figure 11.11: Bifurcation diagrams for the open-loop system with $\epsilon = 0.9$ and $\beta = 1.42$.

where $\kappa = \left(\dfrac{\bar{\beta}}{\beta}\right)^2$ and

$$a_1(R) = -\sigma R \frac{\partial \mathcal{F}(R,\Phi)}{\partial \Phi}\bigg|_{\Phi = \Phi_{R+}(R)} \tag{11.63}$$

$$a_2(R) = -\frac{\partial \mathcal{G}(R,\Phi)}{\partial \Phi}\bigg|_{\Phi = \Phi_{R+}(R)} \tag{11.64}$$

$$\Sigma_1(R) = -\frac{\dfrac{\partial \mathcal{F}(R,\Phi)}{\partial R}}{\dfrac{\partial \mathcal{F}(R,\Phi)}{\partial \Phi}}\bigg|_{\Phi = \Phi_{R+}(R)}$$

$$= \frac{d\Phi_{R+}(R)}{dR} \tag{11.65}$$

$$\Sigma_2(R) = \frac{\partial \mathcal{G}(R,\Phi)}{\partial R}\bigg|_{\Phi = \Phi_{R+}(R)} + \frac{\partial \mathcal{G}(R,\Phi)}{\partial \Phi}\bigg|_{\Phi = \Phi_{R+}(R)} \frac{d\Phi_{R+}(R)}{dR}$$

$$= \frac{\partial \mathcal{G}(R,\Phi)}{\partial R}\bigg|_{\Phi = \Phi_{R+}(R)} - a_2(R)\Sigma_1(R)$$

$$= \frac{d\Psi_{R+}(R)}{dR}. \tag{11.66}$$

Before we derive our stabilization criteria, we establish some fundamental properties of the functions $a_1(R), a_2(R), \Sigma_1(R)$, and $\Sigma_2(R)$. As everything else in this chapter, these properties are independent of a specific form of the

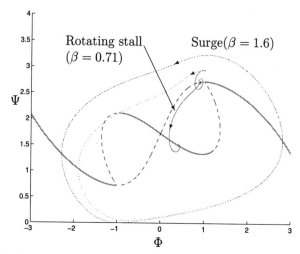

Figure 11.12: Transient responses for throttle opening $\gamma = 1.15$, slightly below the value for the stall inception point. A low value of β ($\beta = 0.71$) results in rotating stall, while a high value of β ($\beta = 1.6$) results in surge.

compressor characteristic.

Lemma 11.1 *For a compressor characteristic that has a maximum at $\Phi = 1$, that is, with $\Psi'_C(1) = 0$ and $\Psi''_C(1) < 0$, we have the following properties for the functions $a_1(R)$, $a_2(R)$, $\Sigma_1(R)$, and $\Sigma_2(R)$ at the stall inception point:*

$$a_1(0) = 0 \tag{11.67}$$

$$a_2(0) = 0 \tag{11.68}$$

$$\Sigma_1(0) = S_1 = -\frac{1}{2}\frac{\Psi'''_C(1)}{\Psi''_C(1)} \tag{11.69}$$

$$\Sigma_2(0) = S_2 = \Psi''_C(1) \tag{11.70}$$

$$a'_1(0) = -\frac{2\sigma}{3}S_2 > 0 \tag{11.71}$$

$$a'_2(0) = S_1 S_2 \tag{11.72}$$

$$\Sigma'_1(0) = \frac{5}{4}S_1^3 - \frac{1}{4S_2}\left(3S_1\Psi^{(4)}_C(1) + \frac{5}{12}\Psi^{(5)}_C(1)\right) \tag{11.73}$$

$$\Sigma'_2(0) = -3S_1^2 S_2 + \frac{1}{2}\Psi^{(4)}_C(1) \tag{11.74}$$

Proof. See Appendix F.

Now we turn our attention to the stability conditions for the linearized model (11.60)–(11.62). Substituting (11.61) into (11.62), we get the Jacobian of the system (11.45)–(11.47):

$$\begin{bmatrix} a_1\Sigma_1 & -a_1 & 0 \\ \Sigma_2 + a_2\Sigma_1 & -a_2 & -1 \\ \kappa\left[(\Sigma_2 + a_2\Sigma_1)\,d_\Phi - c_R\right] & \kappa\left(c_* - a_2 d_\Phi\right) & -\kappa\left(d_\Phi + c_\Psi\right) \end{bmatrix}. \qquad (11.75)$$

The characteristic polynomial of the Jacobian (11.75) is

$$\begin{aligned} p(s) \;=\; & s^3 + \left[\kappa\left(d_\Phi + c_\Psi\right) + \left(a_2 - a_1\Sigma_1\right)\right]s^2 \\ & + \left[\kappa c_* + \kappa\left(a_2 - a_1\Sigma_1\right)c_\Psi + a_1\left(\Sigma_2 - \kappa\Sigma_1 d_\Phi\right)\right]s \\ & + \kappa a_1\left(c_R - \Sigma_1 c_* + \Sigma_2 c_\Psi\right). \end{aligned} \qquad (11.76)$$

By applying the Routh–Hurwitz method, the necessary and sufficient conditions for stability of the system (11.60)–(11.62) are

$$\kappa\left(d_\Phi + c_\Psi\right) + \left(a_2 - a_1\Sigma_1\right) \;>\; 0 \qquad (11.77)$$

$$\kappa c_* + \kappa\left(a_2 - a_1\Sigma_1\right)c_\Psi + a_1\left(\Sigma_2 - \kappa\Sigma_1 d_\Phi\right) \;>\; 0 \qquad (11.78)$$

$$\kappa a_1\left(c_R - \Sigma_1 c_* + \Sigma_2 c_\Psi\right) \;>\; 0 \qquad (11.79)$$

$$\begin{aligned} \left[\kappa\left(d_\Phi + c_\Psi\right) + \left(a_2 - a_1\Sigma_1\right)\right]\left[\kappa c_* + \kappa\left(a_2 - a_1\Sigma_1\right)c_\Psi \right. \\ \left. + a_1\left(\Sigma_2 - \kappa\Sigma_1 d_\Phi\right)\right] - \left[\kappa a_1\left(c_R - \Sigma_1 c_* + \Sigma_2 c_\Psi\right)\right] \;>\; 0 \end{aligned} \qquad (11.80)$$

Since, by Lemma 11.1,

$$a_2\left(R\right) - a_1\left(R\right)\Sigma_1\left(R\right) = O\left(R\right), \qquad (11.81)$$

then

$$c_\Psi + d_\Phi > 0 \qquad (11.82)$$

guarantees that (11.77) is satisfied near the stall inception point. Noting that

$$a_1(R) = a_1'(0)R + O(R^2), \qquad (11.83)$$

and that

$$\Sigma_1(R) \;=\; S_1 + O(R) \qquad (11.84)$$

$$\Sigma_2(R) \;=\; S_2 + O(R), \qquad (11.85)$$

we conclude that the condition

$$c_* > 0 \qquad (11.86)$$

guarantees that condition (11.78) is satisfied near the stall inception point. Furthermore, since $a_1'(0) > 0$, the condition

$$c_R - S_1 c_* + S_2 c_\Psi > 0 \qquad (11.87)$$

guarantees that condition (11.79) is satisfied near the stall inception point. Noting that the expression in (11.80) is $\kappa^2(d_\Phi + c_\Psi)c_* + O(R)$, we conclude that it is satisfied near the stall inception point whenever (11.82) and (11.86) are satisfied. We point out that condition (11.87) coincides with the bifurcation condition (11.52), and (11.86) coincides with (11.56). The stability conditions are summarized in Table. 11.3.

Table 11.3: The stability conditions for the system (11.45)–(11.47).

$c_R - S_1\left(c_\Phi + \dfrac{1}{\beta^2}\right) + S_2 c_\Psi > 0$
$c_\Phi + \dfrac{1}{\beta^2} > 0$
$c_\Psi + d_\Phi > 0$

11.6.3 Stability at the Bifurcation Point

Our analysis showed that we can stabilize stall equilibria *near* the stall inception point but it did not include the stall inception point itself because $a_1(0) = 0$ implies that the characteristic polynomial (11.76) has one root at $s = 0$. For the stall inception point, the analysis based on linearization is inconclusive. Therefore, at this point we apply a center manifold technique.

The one-dimensional center manifold of the equilibrium $R = 0$, $\Phi = 1$, $\Psi = \Psi_C(1)$ of the system (11.45)–(11.47) is readily shown to belong to

$$\Phi(R) = 1 + \frac{c_R + S_2 c_\Psi}{c_*} R + O\left(R^2\right) \qquad (11.88)$$

$$\Psi(R) = \Psi_C(1) + S_2 R + O(R^2). \qquad (11.89)$$

Then its reduced system becomes

$$\dot{R} = \frac{2}{3}\sigma S_2 \left[\frac{c_R - S_1 c_* + S_2 c_\Psi}{c_*} + O(R)\right] R^2. \qquad (11.90)$$

Since $\Psi_C''(1) = S_2 < 0$, the system (11.90) is asymptotically stable if the stability conditions of Table 11.3 are satisfied. Therefore, by the reduction principle [64, Theorem 4.2], the equilibrium $R = 0$, $\Phi = 1$, $\Psi = \Psi_C(1)$ of (11.45)–(11.47) is asymptotically stable.

11.7 Pressure Peak Seeking for the Surge Model

For clarity of presentation, we first consider the model (11.15)–(11.17) restricted to the invariant manifold $R = 0$:

$$\dot{\Phi} = -\Psi + \Psi_C(\Phi) \qquad (11.91)$$

$$\dot{\Psi} = \frac{1}{\beta^2}(\Phi + \Phi_T). \qquad (11.92)$$

This model is, in fact, the surge model introduced by Greitzer [47], which describes limit cycle dynamics in centrifugal compressors. The function $\Psi_C(\Phi)$

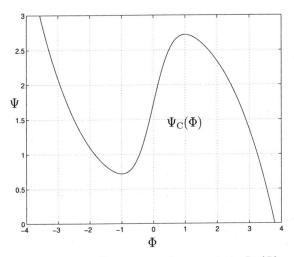

Figure 11.13: Compressor characteristic $\Psi_C(\Phi)$.

represents the "compressor characteristic," whose typical S-shape is given in Figure 11.13. Our objective is to converge to the peak of this characteristic and operate the compressor with maximum pressure. Thus, we denote $x = (\Phi, \Psi)$ and $y = \Psi$. In [68] it was showed that a control law of the form

$$\gamma = \frac{\Gamma + \bar{\beta}^2 \left(c_\Psi \Psi - c_\Phi \Phi\right)}{\sqrt{\Psi}} \tag{11.93}$$

stabilizes equilibria parameterized by Γ. If the design parameters are chosen to satisfy $\bar{\beta}^2 > \beta$, $c_\Psi > 0$, $c_* = c_\Phi + \frac{1}{\bar{\beta}^2} > 0$, and

$$\frac{c_*}{c_\Psi} > \max_\Phi \frac{d\Psi_C(\Phi)}{d\Phi} \tag{11.94}$$

(which is finite), then the control law (11.93) achieves *global exponential stability* of equilibria parameterized by Γ.

To apply the peak seeking scheme, we first check that all three assumptions from Section 5.1 are satisfied:

1. From (11.18), (11.91)–(11.93), Γ is given by

$$\Gamma = \Gamma_\Phi\left(\Phi_0\right) = 1 + \Phi_{C0} + \bar{\beta}^2 \left(c_* \Phi_0 - c_\Psi \Psi_C\left(\Phi_0\right)\right), \tag{11.95}$$

where Φ_0 is the equilibrium value of Φ. In view of (11.94), it is clear that the function $\Gamma_\Phi\left(\cdot\right)$ is invertible. Thus, for each value of Γ, there is only one equilibrium $(\Phi, \Psi) = \left(\Gamma_\Phi^{-1}\left(\Gamma\right), \Psi_C\left(\Gamma_\Phi^{-1}\left(\Gamma\right)\right)\right)$, which means that Assumption 5.1 is satisfied.

2. As we indicated above, it was proved in [68] that (11.94) guarantees that the equilibrium is exponentially stable not only locally but also globally, hence Assumption 5.2 is satisfied.

3. Following the notation in Section 5.1, $y = h \circ l(\Gamma) \triangleq \Psi_C \left(\Gamma_\Phi^{-1}(\Gamma) \right)$. The Moore-Greitzer model (11.15)–(11.17) is scaled so that $\Psi_C(\Phi)$ always has a maximum at $\Phi = 1$. Since $\Gamma_\Phi(\cdot)$ is bijective, $\Psi_C \left(\Gamma_\Phi^{-1}(\Gamma) \right)$ has a maximum at

$$\Gamma^* = 1 + \Phi_{C0} + \bar\beta^2 \left(c_* - c_\Psi \Psi_C(1) \right) , \tag{11.96}$$

that is,

$$\left(\Psi_C \circ \Gamma_\Phi^{-1} \right)'(\Gamma^*) = 0 \tag{11.97}$$

$$\left(\Psi_C \circ \Gamma_\Phi^{-1} \right)''(\Gamma^*) < 0 . \tag{11.98}$$

Hence, Assumption 5.3 is satisfied.

Since Assumptions 5.1–5.3 are satisfied, we can apply the peak seeking scheme given in Figure 11.14 with

$$\Gamma = \hat\Gamma + a \sin \omega t . \tag{11.99}$$

By Theorem 5.5, for sufficiently small ω, δ, and a, $\Phi(t)$ converges to an $O(\omega + \delta + a)$-neighborhood of 1 and $\Psi(t)$ converges to an $O(\omega + \delta + a)$-neighborhood of its maximum value $\Psi_C(1)$.

The application of the peak seeking scheme to the surge model (11.91), (11.92) makes clearer its application to the full Moore-Greitzer model (11.15)–(11.17) in the next section.

11.8   Peak Seeking for the Full Moore-Greitzer Model

Now we consider the full Moore-Greitzer model (11.15)–(11.17). In Section 11.6 we studied the stabilization of this model using the family of control laws

$$\gamma = \frac{\Gamma + \bar\beta^2 \left(c_\Psi \Psi - c_\Phi \Phi + c_R R - d_\Phi \dot\Phi \right)}{\sqrt\Psi} , \tag{11.100}$$

and gave conditions for selecting the parameters c_Ψ, c_Φ, c_R, and d_Φ, such that local stabilization near the peak of the compressor characteristic is achieved. Figure 11.14 shows bifurcation diagrams with respect to the parameter Γ, for the control law (11.100) with $d_\Phi = 0$, applied to the MG model with the compressor characteristic

$$\Psi_C(\Phi) = \Psi_{C0} + 1 + (1 - \epsilon) \left(\frac{3}{2}\Phi - \frac{1}{2}\Phi^3 \right) + \epsilon \frac{2\Phi}{1 + \Phi^2} . \tag{11.101}$$

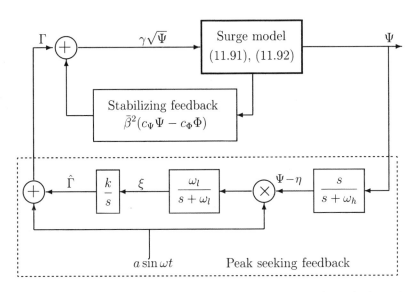

Figure 11.14: Peak seeking scheme for the surge model (11.91), (11.92).

Treating (R, Φ, Ψ) as the state x, the equilibrium map $x = l(\Gamma)$ is given by the bifurcation diagrams in Figure 11.15. The solid curves represent stabilized equilibria, while the dashed curves represent unstable equilibria. As in Section 11.7, we take $y = \Psi$ as the output that we want to maximize using the peak seeking scheme. Unfortunately, none of the three assumptions in Section 5.2 is satisfied: (1) the function $x = l(\Gamma)$ is multi-valued, (2) the equilibrium at the peak is asymptotically stable but not exponentially stable, and (3) the output equilibrium map $y = h \circ l(\Gamma)$ has a maximum but it is not necessarily continuously differentiable. In spite of violating the assumptions, the peak seeking scheme can be applied to the full MG model. The scheme is given in Figure 11.16. The closed loop system is

$$\dot{R} = \frac{\sigma}{3\pi}\sqrt{R}\int_0^{2\pi}\Psi_C\left(\Phi + 2\sqrt{R}\sin\lambda\right)\sin\lambda d\lambda, \quad R(0) \geq 0 \quad (11.102)$$

$$\dot{\Phi} = -\Psi + \frac{1}{2\pi}\int_0^{2\pi}\Psi_C\left(\Phi + 2\sqrt{R}\sin\lambda\right)d\lambda \quad (11.103)$$

$$\dot{\Psi} = \frac{1}{\beta^2}\Big[1 + \Phi_{C0} - \hat{\Gamma} - a\sin\omega t$$
$$-\bar{\beta}^2\left(c_\Psi\Psi - c_\Phi\Phi + c_R R - d_\Phi\dot{\Phi}\right)\Big] \quad (11.104)$$

$$\dot{\hat{\Gamma}} = k\xi \quad (11.105)$$

$$\dot{\xi} = -\omega_l\xi + \omega_l\left(\Psi - \eta\right)a\sin\omega t \quad (11.106)$$

$$\dot{\eta} = -\omega_h\eta + \omega_h y. \quad (11.107)$$

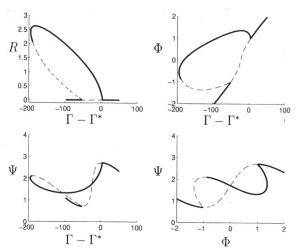

Figure 11.15: Bifurcation diagrams for the case $\epsilon = 0.9$, $\beta = 1.42$, with the full-state controller. The control gains are $c_R = 30$, $c_\Psi = 7$ and $c_\Phi = 20$.

The multiple equilibria in the full MG model make the closed-loop system (11.102)–(11.107) considerably more difficult to analyze than the closed-loop system in the surge model. The *solid* curve in the Ψ vs. $\Gamma - \Gamma^*$ plot in Figure 11.15 plays the role of the function $\nu(\cdot)$ in the averaging analysis. Even though ν is not continuously differentiable, we can show that the average equilibrium is $O(a)$-close to the point $(R, \Phi, \Psi) = (0, 1, \Psi_C(1))$, that it is to the right of the peak (on the flat side of the solid curve on the Ψ vs. $\Gamma - \Gamma^*$ plot), and that it is exponentially stable. The singular perturbation analysis consists of a study of the reduced model and the boundary layer model. The averaging analysis establishes the existence of an exponentially stable $O(\omega+a)$-small periodic orbit of the reduced model—a conclusion no different than for the surge model. The difference comes in the analysis of the boundary layer model:

$$\frac{dR_b}{dt} = \frac{\sigma}{3\pi}\sqrt{R_b}\int_0^{2\pi}\Psi_C\left(\Phi_b + 2\sqrt{R_b}\sin\lambda\right)\sin\lambda\,d\lambda\,,$$
$$R(0) \geq 0 \tag{11.108}$$
$$\frac{d\Phi_b}{dt} = -\Psi_b + \frac{1}{2\pi}\int_0^{2\pi}\Psi_C\left(\Phi_b + 2\sqrt{R_b}\sin\lambda\right)d\lambda \tag{11.109}$$
$$\frac{d\Psi_b}{dt} = \frac{1}{\beta^2}\Big[1 + \Phi_{C0} - \Gamma - a\sin\omega t$$
$$-\bar{\beta}^2\left(c_\Psi\Psi_b - c_\Phi\Phi_b + c_R R_b - d_\Phi\frac{d\Phi_b}{dt}\right)\Big]\,. \tag{11.110}$$

Except for the equilibrium for $\Gamma = \Gamma^*$, the equilibria of interest in the boundary

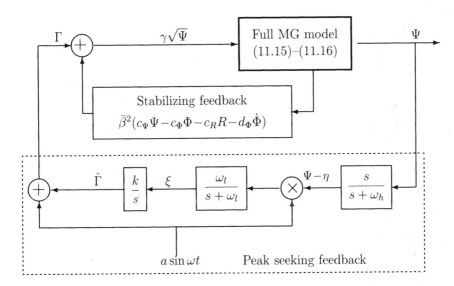

Figure 11.16: Peak seeking scheme for the full Moore-Greitzer model (11.15)–(11.17).

layer model are *exponentially* stable. The equilibrium for $\Gamma = \Gamma^*$ is only *asymptotically* stable. However any ball (with arbitrarily small but nonzero radius) around this equilibrium is exponentially stable. Or, to put it differently, the equilibrium for $\Gamma = \Gamma^*$ is exponentially *practically* stable with an arbitrarily small residual set. This set can be selected to be $O(\omega + a + \delta)$-small. Then by invoking Tikhonov's theorem on the infinite interval, we can draw the same conclusions as in Theorem 5.5, namely, that, for sufficiently small ω, δ, and a, the solution $\big(R(t),\ \Phi(t),\ \Psi(t),\ \hat{\Gamma}(t),\ \xi(t),\ \eta(t)\big)$ converges to an $O(\omega + a + \delta)$-neighborhood of the point $(0,\ 1,\ \Psi_C(1),\ \Gamma^*,\ 0,\ \Psi_C(1))$.

11.9 Simulations for the Full MG Model

We now present simulations of the peak seeking scheme from Figure 11.16 for a compressor with $\Phi_{C0} = 0$, $\Psi_{C0} = 0.72$, $\beta = 1.42$, $\epsilon = 0.9$, $\sigma = 4$, and with a stabilizing controller whose parameters are $\bar{\beta} = 1.42$, $c_\Psi = 2$, $c_\Phi = 4$, $c_R = 7$, $d_\Phi = 0$.

Our first simulation employs a peak seeking scheme with $a = 0.05$, $\omega = 0.03$, $\omega_h = 0.03$, $\omega_l = 0.01$, and $k = 0.4$. The trajectory is shown in Figure 11.17, where the darker curve represents the trajectory and the lighter lines represent the axisymmetric and stall characteristics. The trajectory starts from an equilibrium on the stall characteristic and converges to a small periodic orbit near the peak of the compressor characteristic. If the peak seeking

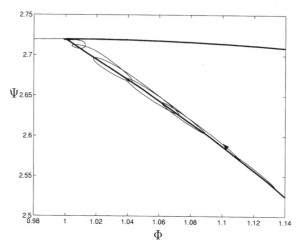

Figure 11.17: Trajectory under the peak seeking feedback for $\omega = 0.03$ and $k = 0.4$.

parameters are selected differently, for example, if k increased to $k = 1.5$, the new shape of the trajectory is shown in Figure 11.18. In this case the convergence is smoother and faster but the periodic orbit is farther from the peak. (Note, however, that, given more time, the periodic orbit would slowly approach the peak.) Even though faster adaptation throws the periodic orbit further to the right of the peak, the periodic orbit remains in the "flat" region of the compressor characteristic where variations in mass flow Φ result in only minor variations of the pressure rise Ψ.

As explained in the section 11.8, the convergence of the trajectory to a close neighborhood of the peak is the result of regulating $\hat{\Gamma}(t)$ to a neighborhood of Γ^*. Figure 11.19 shows the time traces of $\hat{\Gamma}(t) - \hat{\Gamma}(0)$ for the trajectories in Figures 11.17 and 11.18. In this case $\hat{\Gamma}(0) = -1.2$ and $\Gamma^* = -0.90$, which means that $\Gamma^* - \hat{\Gamma}(0) = 0.30$. This explains why the trajectory in Figure 11.17 converges closer to the peak than that in Figure 11.18.

Since a permanent presence of the periodic perturbation is undesirable, we now show that it can be disconnected after a short peak seeking period. Figure 11.20 shows the pressure transient and steady state up to $t = 2550$, at which time, k and a are set to zero. The disconnection of adaptation makes the trajectory transition from the periodic orbit to an equilibrium on the compressor characteristic, just to the right of the peak. To make the transient more visible, we have set $R(2550+0) = 0.06$ because $R(2550-0)$ is practically zero. Figure 11.20 also shows that the pressure variations in steady state under the peak seeking feedback are hardly noticeable, especially if compared to a large gain in the DC value of the pressure. To show the pressure variations more clearly, we zoom in on them in Figure 11.21.

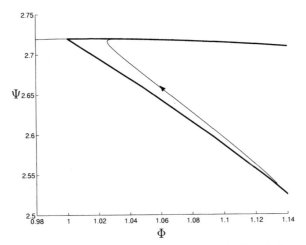

Figure 11.18: Trajectory under the peak seeking feedback for $\omega = 0.03$ and $k = 1.5$.

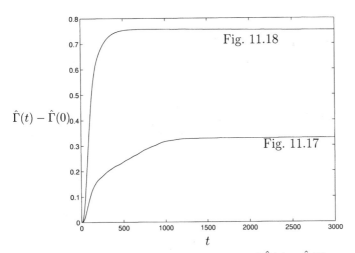

Figure 11.19: Time response of $\hat{\Gamma}(t) - \hat{\Gamma}(0)$.

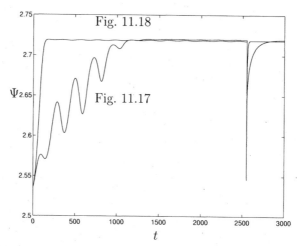

Figure 11.20: Time response of the pressure rise. At $t = 2550$ we disconnect the peak seeking feedback.

Notes and References

The new parameterized model derived here captures the right-skew property of [102] while retaining the relative simplicity of the Moore-Greitzer model [84]. This model can now be easily used in simulations as a replacement of the cubic MG3. The controllers developed are applicable to any Moore-Greitzer type model with an arbitrary $\Psi_C(\Phi)$, though we evaluate them upon the ϵ-MG3 model. The difficulties for control increase with the increase of ϵ.

The idea to use extremum control for maximizing the pressure rise in an aeroengine compressor is not new. As far back as in 1957, George Vasu of the NACA (now NASA) Lewis Laboratory published his experiments in which he varied the fuel flow to achieve maximum pressure [112]. While his engine was apparently not of the kind that could enter either rotating stall or surge instabilities (so local stabilizing feedback was not necessary), it is remarkable that he recognized the opportunity to maximize the pressure by extremum seeking feedback long before the compressor models of the 1970's and 1980's have emerged and the dynamics of compression systems have been understood.

Several feedback designs to stabilize surge and stall through varying throttle opening γ have appeared in the literature beginning with [75], who developed a *local* bifurcation-based controller that changes the character of the bifurcation at the stall inception point, from hard subcritical to soft supercritical, thus avoiding an abrupt transition into rotating stall. The paper [8] experimentally validated this design on a low-speed compressor and [38] developed an improved version of this design which prevents surge on high-speed compressors. The result in [68] reduced the sensing requirement for global

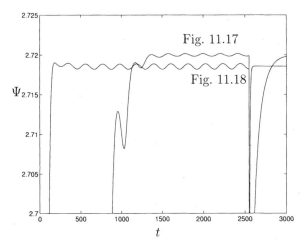

Figure 11.21: Time response of the pressure rise on a scale where the variations are visible.

asymptotic stabilization (GAS) from three to two (Ψ and Φ) measurements but is restricted to *cubic* compressor characteristics proposed in [84]. The design of controllers for compressors with general characteristics is proposed in [13]. Regions of attraction provided by linear and nonlinear controllers were compared in [40]. This chapter draws upon [116] for modeling, and on [117] for application of extremum seeking. More recently, in [58], an output feedback controller achieves semiglobal stabilization using only pressure sensing, and most recently, the result in [4] achieves GAS using only pressure (Ψ) measurement. The result in [118] derives geometric sufficient conditions for stabilization of the bifurcations for use in low spatial actuation authority schemes.

Chapter 12

Compressor Instabilities: Part II

This chapter presents experimental application of extremum seeking feedback to optimize compressor operation. The experiments were performed on an axial-flow compressor in Richard Murray's laboratory at the California Institute of Technology. A prerequisite for experimental validation of the scheme from Section 11.7 is the availability of a high-bandwidth bleed valve for stabilization of rotating stall. However, as shown in [19], rotating stall can also be stabilized by air injection. In this chapter we combine the air injection rotating stall controller from [19] with the extremum seeking scheme from Section 11.7 to achieve maximization of pressure rise. The extremum seeking is implemented via a slow bleed valve, while rotating stall stabilization is performed with air injection, as in [33]. The results demonstrate the effectiveness of extremum seeking in maintaining maximal pressure rise while preventing rotating stall (of either large or small amplitude).

In this chapter, we also apply slope seeking feedback in simulation to the ϵ-MG3 of compressor surge and stall presented in Chapter 11 and demonstrate: (1) near-optimal compressor operation with only pressure sensing; and (2) robustness of the control to finite disturbances.

The laboratory implementation of extremum seeking on an axial-flow compressor is described in Section 12.1. The main experimental results are shown in Section 12.2 and Section 12.3 illustrates near optimal compressor operation under slope seeking feedback.

12.1 Extremum Seeking on the Caltech Rig

Experimental Setup. The Caltech compressor rig is a single-stage, low-speed, axial compressor with sensing and actuation capabilities. Figure 12.1 shows a magnified view of the sensor and injection actuator ring and Figure 12.2 a drawing of the rig. The compressor is an Able Corporation model

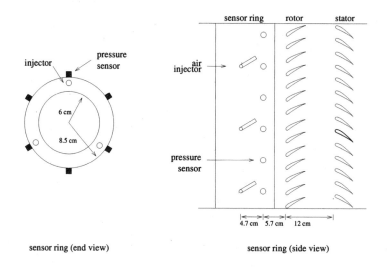

Figure 12.1: Sensor and injection actuator ring.

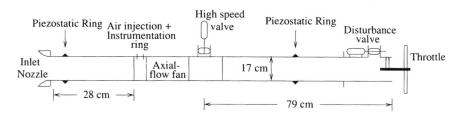

Figure 12.2: Experimental setup.

29680 single-stage axial compressor with 14 blades, a tip radius of 8.5 cm, and a hub radius of 6 cm. The blade stagger angle varies from 30° at the tip to 51.6° at the hub, and the rotor to stator distance is approximately 12 cm. Experiments are run with a rotor frequency of 100 Hz, giving a tip Mach number of 0.17. Rotating stall is observed under this condition on the Caltech rig with a frequency of 65 Hz while surge occurs at approximately 1.7 Hz. Data taken for a stall transition event suggests that the stall cell grows from the noise level to its fully developed size in approximately 30 msec (three rotor revolutions). At the stall inception point, the velocity of the mean flow through the compressor is approximately 16 m/sec.

Six static pressure transducers with 1000 Hz bandwidth are evenly distributed along the annulus of the compressor at approximately 5.7 cm from the rotor face. A discrete Fourier transform is performed on the signals from the transducers, and the amplitude and phase of the first and second mode of the stall cell associated with the total-to-static pressure perturbation are obtained. The difference between the pressure obtained from one static pressure transducer mounted at the piezostatic ring at the inlet and that from one mounted at another piezostatic ring downstream near the outlet of the system is computed as the pressure rise across the compressor. All of the static pressure transducer signals are filtered through a fourth order Bessel low pass filter with a cutoff frequency of 1000 Hz before the signal processing phase in the software.

The air injectors are on-off type injectors driven by solenoid valves. For applications on the Caltech compressor rig, the injectors are fed with a pressure source supplying air at a maximum pressure of 80 psi. Due to significant losses across the solenoid valves and between the valves and the pressure source, the injector back pressure reading does not represent an accurate indication of the actual velocity of the injected air on the rotor face. For example, using a hotwire anemometer, the maximum velocities of the injected air measured at a distance equivalent to the rotor-injector distance for 50 and 60 psi injector back pressure are approximately 30.2 and 33.8 m/sec respectively. At the stall inception point, each injector can add approximately 1.7% mass, 2.4% momentum, and 1.3% energy to the system when turned on continuously at 60 psi injector back pressure. The bandwidth associated with the injectors is approximately 200 Hz at 50% duty cycle. Extremum seeking is implemented via a bleed valve whose bandwidth is 51 Hz.

12.1.1   Actuation for Stall Stabilization

A prerequisite for experimental validation of the scheme from Section 11.7 is the availability of a high-bandwidth bleed valve for stabilization of rotating stall. However, as shown in [33], rotating stall can also be stabilized by air injection. In this section we combine the air injection rotating stall controller

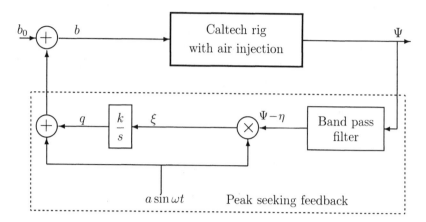

Figure 12.3: Peak seeking scheme for the Caltech rig.

from [33] with the extremum seeking scheme from Section 11.7 to achieve maximization of pressure rise.

Stall stabilization is performed by air injection in a one-dimensional on-off fashion. When the measured amplitude of the first mode of the stall cell is above a certain threshold, all three injectors are fully open. Otherwise they are closed.

The set point of the compressor is varied by a bleed valve. The characteristic of the pressure rise Ψ with respect to the bleed angle b is shown in the Figure 12.4. Note that higher bleed angle means lower overall throttle opening. There is a clear peak in the characteristic curve, so we can apply the extremum seeking to regulate the system to the bleed set point that maximizes the pressure rise. The points to the left of the peak are axisymmetric equilibria. The points to the right of the peak are stabilized low amplitude stall equilibria.

12.1.2　Filter Design

In the theoretical analysis presented in Chapter 11 noise was not considered and a high-pass filter $\dfrac{s}{s + \omega_h}$ was employed. Because of noise in the experiment we use a band pass filter. From the power spectrum analysis, we learn that the noise is above 150 Hz. We also know that the stall frequency is about 65 Hz. Since the filter should cut out both the high frequency noise and the stall oscillations, as well as DC, we choose the pass band to be 4 to 6 Hz, and implement it as a third order Butterworth filter.

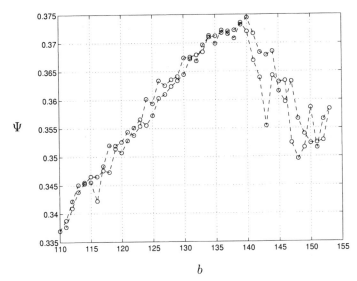

Figure 12.4: Ψ vs. bleed angle b.

12.2   Experimental Results

12.2.1   Initial Point on the Axisymmetric Characteristic

We select the integrator gain as 600 and set the frequency of the perturbation to 5 Hz. From Figure 12.4 we know that the peak is around 134–139°. We set the initial bleed angle at 110°, the farthest point to the left in Figure 12.4, and set the perturbation to 3°. The perturbed bleed angle is shown in Figure 12.5 and the pressure rise response is shown in Figure 12.6. Comparing the peak pressure rise 0.372 of Figure 12.4 to that of Figure 12.6, we see that they are the same. Note that the directions of b on this figure (Figure 12.4) and Φ in Figure 11.1 are opposite. In this figure, the no-stall (axisymmetric) equilibria are to the left of the peak, while the ones to the right are the stall equilibria (low stall, with air injection controller).

12.2.2   Initial Point on the Nonaxisymmetric Characteristic

For the compressor control, the most important issue is to control the system to avoid the stall that causes the deep pressure drop. Since the air injection can control stall for a reasonable interval, we can set the initial point at the stall characteristic. We select the initial point as 150° bleed valve angle because from Figure 12.4 we can see that 150° is the largest angle we can achieve without a deep drop of the pressure rise. In this case we choose the gain of the

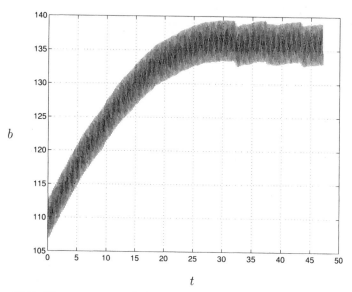

Figure 12.5: Time response of the bleed angle initiating from the *axisymmetric* characteristic.

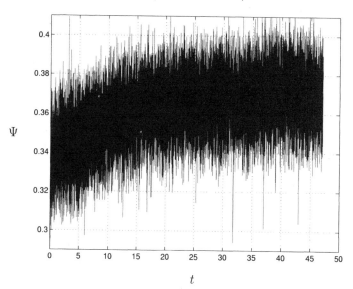

Figure 12.6: Time response of the pressure rise initiating from the *axisymmetric* characteristic.

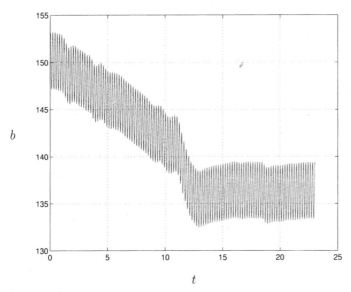

Figure 12.7: Time response of the bleed angle initiating from the *stall* characteristic.

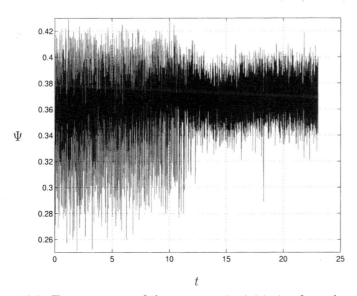

Figure 12.8: Time response of the pressure rise initiating from the *stall* characteristic.

integrator as 400. The perturbation signal is set to $3°$ as in the axisymmetric case. The perturbed bleed angle is shown in Figure 12.7 and the pressure rise response is shown in Figure 12.8. The peak pressure in Figure 12.8 coincides with that in Figure 12.4.

A closer look at Figures 12.7 and 12.8 shows that the rotating stall amplitude is reduced as the bleed angle is reduced. The effect of extremum seeking on a system starting in rotating stall is to bring it out of stall without pushing the operating point away from the peak and reducing the pressure rise.

In both Figures 12.5 and 12.7 one can observe fluctuations of the mean of the bleed angle at the peak. Comparing with Figures 12.6 and 12.8, we see that these fluctuations occur at the same time when pressure drops resembling stall inception occur. Extremum seeking reacts to this by pushing the operating point further to the right on the axisymmetric characteristic and then slowly returning it to the peak.

12.3   Near Optimal Compressor Operation via Slope Seeking

For the purpose of our study, we consider a three-stage compressor considered in [116] with parameters $\Psi_{C0} = 0.72$, $\Phi_{C0} = 0$ and $\sigma = 4$. Furthermore we choose the low speed case of $\beta = 0.71$ from [116]. Figures 12.9 and 12.10 show the bifurcation diagrams for the compressor for $\epsilon = 0$ and $\epsilon = 0.9$ respectively; the solid lines showing stable equilibria and the dashed lines showing unstable equilibria.

The performance objective for compressors is to maximize the pressure rise Ψ with respect to the mass flow Φ without entering stall or surge instabilities. But as seen from the Ψ versus γ diagrams in Figures 12.9 and 12.10, the point of maximum pressure rise is directly above a stable stall equilibrium, and the stable high pressure branch ends at the maximum. The stall branch comes very deep under the high pressure branch in the case when $\epsilon = 0.9$, showing a deep hysteresis characteristic of some high performance compressors.

Thus, running the compressor at maximum performance risks entering a stall cycle under any small disturbance. Here, we illustrate achievement of near-optimal performance of the compressor under slope-seeking feedback that uses only the pressure measurement Ψ, and actuates throttle opening γ. Through slope seeking, we can operate at a point on the compressor characteristic that is just short of the maximum. This is done by using a slope setting $r(f'_{ref})$ with commanded slope

$$f'_{ref} = \left(\frac{d\Psi}{d\gamma}\right)_{ref}$$

small and negative in the slope seeking scheme (Figure 3.1).

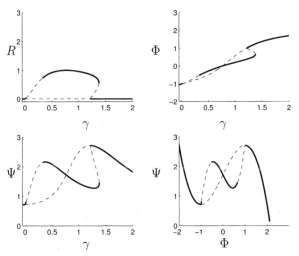

Figure 12.9: Bifurcation diagrams for the open-loop system with $\epsilon = 0$ and $\beta = 0.71$. The throttle opening γ is the bifurcation parameter.

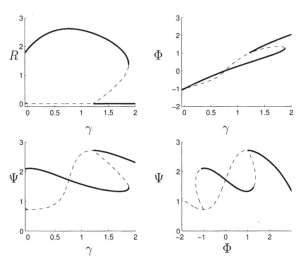

Figure 12.10: Bifurcation diagrams for the open-loop system with $\epsilon = 0.9$ and $\beta = 0.71$. The throttle opening γ is the bifurcation parameter.

Slope Seeking Design. We design two slope seeking loops; one for the case of low hysteresis, $\epsilon = 0$, and for the case of deep hysteresis, $\epsilon = 0.9$. In both designs, we choose forcing frequency $\omega = 0.5$ (this corresponds to an oscillation frequency of about 16 Hz when transformed to normal time units), forcing amplitude $a = 0.025$, gain $k = -0.6$, and pole of washout filter $h = 0.5$. We set commanded slopes $f'_{ref} = -0.9$ and $f'_{ref} = -0.5$ for the $\epsilon = 0$ and $\epsilon = 0.9$ cases respectively obtaining values of $r(f'_{ref}) = -\frac{a f'_{ref}}{2} \operatorname{Re} \left\{ \frac{j\omega}{j\omega + h} \right\} = 0.0056$ and 0.0031 for the slope settings neglecting plant dynamics.

Simulation Results. We perform simulations (in MATLAB and SIMULINK) with low performance initial conditions of $R(0) = 1$, $\Phi(0) = 1.8565$, $\Psi(0) = 1.3055$, $\gamma(0) = 1.5$ for the $\epsilon = 0$ case, and initial conditions of $R(0) = 1$, $\Phi(0) = 1.4877$, $\Psi(0) = 1.9463$, $\gamma(0) = 1.5$ for the $\epsilon = 0.9$ case. Figures 12.11 and 12.12 show the results for slope seeking (solid lines) along with results for extremum seeking, $r = 0$ (in dotted lines) for the initial conditions above. The results reveal the following features:

1. Both slope seeking and extremum seeking converge to their desired set points: extremum seeking to the maximum pressure, and slope seeking to a point slightly below the maximum.

2. Under a small disturbance at $t = 600$, the system with extremum seeking is destabilized and the system goes into the stall regime, while slope seeking feedback recovers its performance.

Notes and References

This chapter is based upon [6, 117]. In [117], extremum seeking feedback was used in an experiment to optimize performance of a compressor stabilized by air-injection; this led to less demanding sensing and actuation requirements than stabilization of stalled equilibria. In [6], slope seeking feedback is applied to compressor optimization; the work uses only a pressure measurement, and does not need high actuator bandwidth as it does not operate at the point of maximum pressure rise, and therefore does not need to stabilize the compressor.

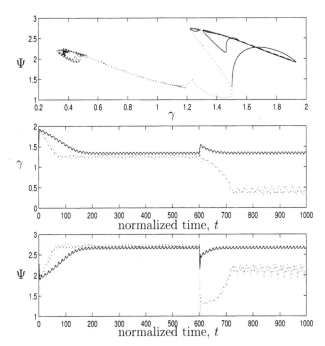

Figure 12.11: Low hysteresis compressor: $\epsilon = 0$

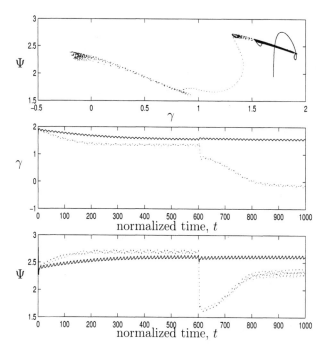

Figure 12.12: Deep hysteresis compressor: $\epsilon = 0.9$

Appendices

Appendix A

Continuous Time Lemmas

Lemma A.1 *If the transfer function $H(s)$ has all of its poles with negative real parts, then for any real ψ,*

$$H(s)\left[\sin(\omega t - \psi)\right] = \mathbf{Im}\left\{H(j\omega)e^{j(\omega t - \psi)}\right\} + \epsilon^{-t}, \tag{A.1}$$

where ϵ^{-t} denotes exponentially decaying terms.

This is simply the frequency response of an asymptotically stable LTI system.

Lemma A.2 *For any two rational functions $A(\cdot)$ and $B(\cdot, \cdot)$, the following is true:*

$$\mathbf{Im}\left\{e^{j(\omega_a t - \psi)}A(j\omega_a)\right\}\mathbf{Im}\left\{e^{j(\omega_b t - \phi)}B(s, j\omega_b)[z(t)]\right\}$$
$$= \frac{1}{2}\mathbf{Re}\left\{e^{j((\omega_b - \omega_a)t + \psi - \phi)}A(-j\omega_a)B(s, j\omega_b)[z(t)]\right\}$$
$$- \frac{1}{2}\mathbf{Re}\left\{e^{j((\omega_b + \omega_a)t - \psi - \phi)}A(j\omega_a)B(s, j\omega_b)[z(t)]\right\}.$$

Proof: Follows by substituting the representations for the real and imaginary parts of a complex number z, $\mathbf{Re}\{z\} = \frac{z + \bar{z}}{2}$, and $\mathbf{Im}\{z\} = \frac{z - \bar{z}}{2}$.

Lemma A.3 (Singular Perturbation) *Consider the singularly perturbed system given by the equations:*

$$\dot{\mathbf{x}} = \mathbf{A}_1(t)\mathbf{x} + \mathbf{B}_1(t)\mathbf{u}, \quad \mathbf{x}(t_0) = \xi(\epsilon) \tag{A.2}$$
$$\mathbf{v} = \mathbf{C}_1(t)\mathbf{x}$$
$$\epsilon\dot{\mathbf{z}} = \mathbf{A}_2\mathbf{z} + \mathbf{B}_2\mathbf{v}, \quad \mathbf{z}(t_0) = \eta(\epsilon) \tag{A.3}$$
$$\mathbf{u} = \mathbf{C}_2\mathbf{z},$$

where $\xi(\epsilon)$ and $\eta(\epsilon)$ are smooth functions of ϵ. If $\mathbf{A}_2$ is Hurwitz and the origin of the reduced LTV model

$$\dot{\bar{\mathbf{x}}} = \left(\mathbf{A}_1(t) + \mathbf{B}_1(t)\mathbf{C}_2\mathbf{A}_2^{-1}\mathbf{B}_2\mathbf{C}_1(t)\right)\bar{\mathbf{x}} \tag{A.4}$$

is exponentially stable, then there exists $\epsilon^* > 0$ such that for all $0 < \epsilon < \epsilon^*$, the system in Eqns. (A.2), (A.3) has a unique solution $\mathbf{x}(t, \epsilon), \mathbf{z}(t, \epsilon)$ defined for all $t \geq t_0 \geq 0$ and $\mathbf{x}(t, \epsilon) - \bar{\mathbf{x}}(t) = \mathrm{O}(\epsilon)$.

Proof: A direct consequence of Theorem 9.4 in Khalil [64].

Appendix B

Discrete Time Lemmas

The following lemmas are used to facilitate the extremum seeking system analysis:

Lemma B.1 *If the transfer function $H(z)$ has all of its poles inside the unit circle and real-valued impulse response, then, for any real ψ,*

$$H(z)\big[\cos(\omega k - \psi)\big] = \mathbf{Re}\big\{H(e^{j\omega})e^{j(\omega k - \psi)}\big\} + \varepsilon^{-k}$$
$$= \big|H(e^{j\omega})\big|\cos(wk - \psi + \psi_H) + \varepsilon^{-k},$$

where $\psi_H = \angle\big(H(e^{j\omega})\big)$.

Lemma B.2 *If the transfer functions $G(z)$ and $H(z)$ have all of their poles inside the unit circle, the following statement is true for any real ϕ and any uniformly bounded $v(k)$:*

$$G(z)\Big[\big(H(z)[\cos(\omega k - \phi)]\big)v(k)\Big] = \mathbf{Re}\Big\{e^{j(\omega k - \phi)}H(e^{j\omega})G(e^{j\omega}z)\big[v(k)\big]\Big\} + \varepsilon^{-k}.$$
(B.1)

Proof. The lemma is proved using the following straightforward calculation:

$$G(z)[(H(z)[\cos(\omega k - \phi)])v(k)]$$
$$= G(z)[\mathbf{Re}\{H(e^{j\omega})e^{j(\omega k - \phi)}\}v(k) + \varepsilon^{-k}] \text{ (by Lemma B.1)} \quad (B.2)$$
$$= \mathbf{Re}\{e^{-j\phi}\mathcal{Z}^{-1}\{G(z)H(e^{j\omega})V(e^{-j\omega}z)\}\} + \varepsilon^{-k} \quad (B.3)$$
$$= \mathbf{Re}\{e^{j(\omega k - \phi)}H(e^{j\omega})\mathcal{Z}^{-1}\{G(e^{j\omega}z)V(z)\}\} + \varepsilon^{-k} \quad (B.4)$$
$$= \mathbf{Re}\{e^{j(\omega k - \phi)}H(e^{j\omega})G(e^{j\omega}z)[v(k)]\} + \varepsilon^{-k}. \quad (B.5)$$

Q.E.D.

Lemma B.3 *For any two rational functions $A(\cdot)$ and $B(\cdot, \cdot)$, the following is true:*

$$\mathbf{Re}\big\{e^{j(\omega k - \psi)} A(e^{j\omega})\big\} \mathbf{Re}\big\{e^{j(\omega k - \phi)} B(z, e^{j\omega})[v(k)]\big\}$$
$$= \frac{1}{2}\mathbf{Re}\big\{e^{j(\psi - \phi)} A(e^{-j\omega}) B(z, e^{j\omega})[v(k)]\big\}$$
$$+ \frac{1}{2}\mathbf{Re}\big\{e^{j(2\omega k - \psi - \phi)} A(e^{j\omega}) B(z, e^{j\omega})[v(k)]\big\}. \tag{B.6}$$

Lemma B.4 *For any rational function $B(\cdot, \cdot)$, the following is true:*

$$\mathbf{Re}\big\{e^{j(\omega k - \phi)} B(z, e^{j\omega})[v(k)]\big\} = \cos(\omega k - \phi)\mathbf{Re}\big\{B(z, e^{j\omega})[v(k)]\big\}$$
$$- \sin(\omega k - \phi)\mathbf{Im}\big\{B(z, e^{j\omega})[v(k)]\big\}. \tag{B.7}$$

Lemma B.5 *Suppose that the transfer functions $H(z)$ and $G(z)$ have all of their poles inside the unit circle, and have minimal state space realizations (A_1, B_1, C_1, D_1) and (A_2, B_2, C_2, D_2) respectively. Then,*

$$G(z)\big[\cos(\omega k - \phi) H(z)[v(k)]\big]$$

can be represented in a state space form as

$$A(k) = \left[\begin{array}{c|c} A_1 & 0 \\ \hline \cos(\omega k) B_2 C_1 & A_2 \end{array} \right],$$
$$B(k) = \left[\begin{array}{c} B_1 \\ \hline \cos(\omega k) B_2 D_1 \end{array} \right]$$
$$C(k) = \left[\begin{array}{c|c} \cos(\omega k) D_2 C_1 & C_2 \end{array} \right]$$
$$D(k) = \cos(\omega k) D_2 D_1,$$

where $A(k)$ is exponentially stable.

Proof. Let $x_1(k)$ and $x_2(k)$ be the state vectors of $H(z)$ and $G(z)$ respectively. Then, $G(z)\big[\cos(\omega k - \phi) H(z)[v(k)]\big]$ is represented in state space form as

$$x_1(k+1) = A_1 x_1(k) + B_1 v(k) \tag{B.8}$$
$$y_1(k) = C_1 x_1(k) + D_1 v(k), \tag{B.9}$$
$$x_2(k+1) = A_2 x_2(k) + B_2 \cos(\omega k - \phi) y_1(k) \tag{B.10}$$
$$y_2(k) = C_2 x_2(k) + D_2 \cos(\omega k - \phi) y_1(k), \tag{B.11}$$

where $y_1(k) = H(z)[v(k)]$ and $y_2(k) = G(z)[\cos(\omega k - \phi) y_1(k)]$. Combining the above two state space forms yields

$$x(k+1) = A(k)x(k) + B(k)v(k) \tag{B.12}$$
$$y_2(k) = C(k)x(k) + D(k)v(k), \tag{B.13}$$

where

$$
\begin{aligned}
x^T(k) &= [x_1^T(k) \mid x_2^T(k)] \\
A(k) &= \left[\begin{array}{c|c} A_1 & 0 \\ \hline c(\omega k)B_2C_1 & A_2 \end{array} \right] \\
B(k) &= \left[\begin{array}{c} B_1 \\ \hline c(\omega k)B_2D_1 \end{array} \right] \\
C(k) &= \left[\begin{array}{cc|c} c(\omega k)D_2 & C_1 & C_2 \end{array} \right] \\
D(k) &= c(\omega k)D_2D_1,
\end{aligned}
$$

and

$$
c(\omega k) \triangleq \cos(\omega k - \phi).
$$

Therefore, $(A(k), B(k), C(k), D(k))$ can be a state space realization of $G(z)\big[\cos(\omega k - \phi)H(z)[v(k)]\big]$.

Since A_1 and A_2 in $A(k)$ are exponentially stable, given any $Q_1 = Q_1^T > 0$ and $Q_2 = Q_2^T > 0$, there exist $P_1 = P_1^T > 0$ and $P_2 = P_2^T > 0$, which are the unique solutions of the following linear equations, respectively:

$$
A_1^T P_1 A_1 - P_1 = -Q_1 \quad \text{and} \quad A_2^T P_2 A_2 - P_2 = -Q_2. \tag{B.14}
$$

By constructing a block diagonal matrix $P = \left[\begin{array}{c|c} P_1 & 0 \\ \hline 0 & P_2 \end{array} \right]$, $P = P^T > 0$, we obtain

$$
A^T(k)PA(k) - P = \left[\begin{array}{c|c} -Q_{11}(k) & R^T(k) \\ \hline R(k) & -Q_2 \end{array} \right], \tag{B.15}
$$

where $Q_{11}(k) = Q_1 - c^2(\omega k)C_1^T B_2^T P_2 B_2 C_1$ and $R(k) = c(\omega k)A_2^T P_2 B_2 C_1$. For any given Q_2, we can choose Q_1 such that

$$
\lambda_{min}(Q_1) > \lambda_{max}\big(c^2(\omega k)C_1^T B_2^T P_2 B_2 C_1\big) + \frac{\lambda_{max}\big(R(k)R^T(k)\big)}{\lambda_{min}(Q_2)} \tag{B.16}
$$

for all $k > 0$. This Q_1 enables $-Q_{11}(k)$ and $-Q_2 + R(k)Q_{11}^{-1}(k)R^T(k)$ to be negative definite for all $k > 0$, and we obtain the following decomposition of Eqn. (B.15):

$$
A^T(k)PA(k) - P = \left[\begin{array}{c|c} I & 0 \\ \hline -R(k)Q_{11}^{-1}(k) & I \end{array} \right] \left[\begin{array}{c|c} -Q_{11}(k) & 0 \\ \hline 0 & \Delta \end{array} \right] \left[\begin{array}{c|c} I & -Q_{11}^{-1}(k)R^T(k) \\ \hline 0 & I \end{array} \right],
$$
$$\tag{B.17}$$

where $\Delta = -Q_2 + R(k)Q_{11}^{-1}(k)R^T(k) < 0$ for all $k > 0$. Consequently, $A^T(k)PA(k) - P < 0$ for all $k > 0$, and $A(k)$ is exponentially stable from the Lyapunov stability theory. Q.E.D.

Appendix C

Aircraft Dynamics in Close Formation Flight

C.1   C-5 and Flight Condition Data

Quantity	Symbol	Measure
Wingspan	b	222 ft, 8 in
Lenght	l	247 ft, 11 in
Height	h	65 ft, 1 in
Wing area	S	6200 ft^2
Root chord	c_R	45 ft, 5 in
Tip chord	c_T	15 ft, 4 in
Aspect ratio	AR	7.3
Taper ratio	TR	0.34
Wing 1/4-chord sweep back	λ	25 deg
Maximum takeoff weight	W_{TOmax}	764,500 lb

Table C.1: C-5 Galaxy Data

C.2   Wake-Induced Velocity Field

The wake-induced upwash distribution is:

$$W_{wake}(x, y, z) = W_R(x, y, z) + W_L(x, y, z), \qquad (C.1)$$

where

$$W_R(x, y, z) = \frac{\Gamma}{4\pi} \frac{y}{y^2 + z^2 + r_c^2} \left(1 + \frac{x}{\sqrt{x^2 + y^2 + z^2}}\right), \qquad (C.2)$$

207

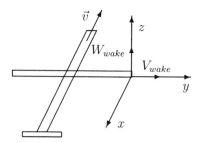

Figure C.1: Wake velocities

and

$$W_L(x, y, z) = \frac{\Gamma}{4\pi} \frac{y + b}{(y + b)^2 + z^2 + r_c^2} \left(1 + \frac{x}{\sqrt{x^2 + (y + b)^2 + z^2}} \right), \qquad \text{(C.3)}$$

are the contributions, respectively, from the right and from the left trailing vortex. The wake-induced sidewash distribution is:

$$V_{wake}(x, y, z) = V_R(x, y, z) + V_L(x, y, z), \qquad \text{(C.4)}$$

where

$$V_R(x, y, z) = \frac{\Gamma}{4\pi} \frac{z}{y^2 + z^2 + r_c^2} \left(1 + \frac{x}{\sqrt{x^2 + y^2 + z^2}} \right), \qquad \text{(C.5)}$$

and

$$V_L(x, y, z) = \frac{\Gamma}{4\pi} \frac{z}{(y + b)^2 + z^2 + r_c^2} \left(1 + \frac{x}{\sqrt{x^2 + (y + b)^2 + z^2}} \right), \qquad \text{(C.6)}$$

are the contributions, respectively, from the right and left trailing vortices. Figure C.1 shows the origin of the coordinate system at the right wing tip of the leader and the directions of the upwash, W_{wake}, and sidewash, V_{wake}.

C.3   Free Flight Model Data

Tables C.2 to C.4 give the states ($\mathbf{x}_{long,8\times1}$, $\mathbf{x}_{lat,8\times1}$) and the inputs of the model, as well as the parameters of the actuators. TE indicates trailing edge. All quantities are to be intended as perturbations from their value at the reference condition for which linearization is performed, the exception being the separations, x, y, and z.

State:	Symbol	Unit	Positive
Ground speed	V	kt	Forward
Vertical velocity	w	ft/s	Down
Pitch rate	q	deg/s	Nose up
Pitch angle	θ	deg	Nose up
Longitudinal separation	x	ft	Behind leader
Vertical separation	z	ft	Above leader
Elevators	δ_e	deg	Elevator TE up
Engines - Thrust	δ_{th}	deg	More thrust
Lateral velocity	v	ft/s	Right wing
Roll Rate	p	deg/s	Right wing down
Yaw rate	r	deg/s	Nose right
Bank angle	φ	deg	Right wing down
Heading	ψ	deg	Nose right
Lateral separation	y	ft	Right of leader
Ailerons	δ_a	deg	Right aileron TE up
Rudder	δ_r	deg	Rudder TE right

Table C.2: States of wingman dynamics

Input: $\mathbf{u}_{clong}$	Symbol	Unit	Positive
Elevators command	δ_{ec}	deg	Elevator TE up
Thrust command	δ_{thc}	deg	More thrust
Ailerons command	δ_{ac}	deg	Right aileron TE up
Rudder command	δ_{rc}	deg	Rudder TE right
Average upwash	W_{wake}	ft/s	Upwards
Rolling moment	L_{wake}	lb*ft	Right wing down
Sidewash	V_{wake}^{CL}	ft/s	To the left

Table C.3: Inputs of wingman dynamics

Control	(Pole) Frequency	Saturation
δ_e	10 rad/s	+25/-25 deg
δ_{th}	0.2 rad/s	+10/-30 klb
δ_a	10 rad/s	+25/-25 deg
δ_r	10 rad/s	+25/-25 deg

Table C.4: Actuation(first order lags): dynamics and saturations

$$
A_{long} = \begin{pmatrix}
-0.00380 & 0.0180 & -0.470 & -0.332 & 0 & 0 & -0.0103 & 0.0000291 \\
-0.102 & -0.427 & 13.0 & -0.0343 & 0 & 0 & 0.286 & 0.00000172 \\
-0.0214 & -0.0963 & -0.645 & 0.000367 & 0 & 0 & 0.938 & 0.00000816 \\
0 & 0 & 1.00 & 0 & 0 & 0 & 0 & 0 \\
1.69 & 0 & 0 & 0 & 0 & 0 & 0 & 0 \\
0 & -0.998 & 0 & 13.0 & 0 & 0 & 0 & 0 \\
0 & 0 & 0 & 0 & 0 & 0 & -10.0 & 0 \\
0 & 0 & 0 & 0 & 0 & 0 & 0 & -0.200
\end{pmatrix}
$$
$$(C.7)$$

$$
B_{long} = \begin{pmatrix}
0 & 0 & 0 & 0 & 0 & 0 & 10.0 & 0 \\
0 & 0 & 0 & 0 & 0 & 0 & 0 & 0.200
\end{pmatrix}^T
\tag{C.8}
$$

$$
A_{lat} = \begin{pmatrix}
-0.0636 & 0.794 & -13.0 & 0.561 & 0 & 0 & -0.000679 & -0.118 \\
-0.0831 & -0.706 & 0.233 & 0 & 0 & 0 & 0.298 & -0.112 \\
0.0182 & -0.0776 & -0.0991 & 0 & 0 & 0 & 0.00618 & 0.324 \\
0 & 1 & 0.0612 & 0 & 0 & 0 & 0 & 0 \\
0 & 0 & 1.00 & 0 & 0 & 0 & 0 & 0 \\
1.00 & 0 & 0 & -0.794 & 13.0 & 0 & 0 & 0 \\
0 & 0 & 0 & 0 & 0 & 0 & -10.0 & 0 \\
0 & 0 & 0 & 0 & 0 & 0 & 0 & -10.0
\end{pmatrix}
$$
$$(C.9)$$

$$
B_{lat} = \begin{pmatrix}
0 & 0 & 0 & 0 & 0 & 0 & 10.0 & 0 \\
0 & 0 & 0 & 0 & 0 & 0 & 0 & 10.0
\end{pmatrix}^T
\tag{C.10}
$$

C.4 Formation Flight Model: Influence Matrices

$$F_W = (\, 0.0180 \quad -0.428 \quad -0.0965 \quad 0 \quad 0 \quad 0 \quad 0 \quad 0 \,)^T \tag{C.11}$$
$$F_L = (\, 0 \quad 0.00000206 \quad 0 \quad 0 \quad 0 \quad 0 \quad 0 \quad 0 \,)^T \tag{C.12}$$
$$F_V = (\, -0.0636 \quad -0.0831 \quad -0.0182 \quad 0 \quad 0 \quad 0 \quad 0 \quad 0 \,)^T \tag{C.13}$$

C.5 Formation-Hold Autopilot Parameters

The classical separations-tracking part of the autopilot is made up by PD compensators:

$$k_{Px} = 0.030, \quad k_{Py} = 12, \quad k_{Pz} = 25 \tag{C.14}$$
$$k_{Dx} = 0.025, \quad k_{Dy} = 0, \quad k_{Dz} = 0 \tag{C.15}$$

and by rate limiters:

$$|V_x|_{max} = 4kt, \quad |V_y|_{max} = 250 ft/min, \quad |V_z|_{max} = 500 ft/min \tag{C.16}$$

The state-space relative-velocities-tracking part of the autopilot is made up by state-proportional gain matrices:

$$K_{x_{long}} = \begin{pmatrix} -57.0 & -909 & 50.1 & 1090 & 0 & 0 & 1.74 & -0.0297 \\ 20400 & -8010 & 411 & 17900 & 0 & 0 & -0.0595 & 1.82 \end{pmatrix}$$
$$\text{(C.17)}$$

$$K_{x_{lat}} = \begin{pmatrix} 629 & 12.3 & 24.7 & -9.39 & 654 & 0 & 0.362 & 0.0540 \\ 289 & 1.76 & 29.2 & -10.4 & 341 & 0 & 0.0540 & 0.557 \end{pmatrix} \quad \text{(C.18)}$$

and by the integrative part whose gain matrices are given by

$$K_{V_x} = \begin{pmatrix} 15.6 \\ -2180 \end{pmatrix}, \quad K_{V_z} = \begin{pmatrix} -1.38 \\ -9.88 \end{pmatrix} \tag{C.19}$$

$$K_{V_y} = \begin{pmatrix} -0.403 \\ -0.193 \end{pmatrix}, \quad K_{\beta} = \begin{pmatrix} -13.6 \\ -28.5 \end{pmatrix} \tag{C.20}$$

Appendix D

Derivation of Equations (11.8) and (11.10)

Using MATHEMATICA, we get

$$
\int_0^{2\pi} \frac{2(\Phi + A\sin\theta)}{1 + (\Phi + A\sin\theta)^2} d\theta
$$
$$
= 2\mathrm{sgn}(\Phi)\left[\left(\Phi^2 - A^2 - 1 - 2i\Phi\right)^{-1/2} + \left(\Phi^2 - A^2 - 1 + 2i\Phi\right)^{-1/2}\right] \quad\text{(D.1)}
$$

Even though (D.1) appears to be a complex expression, we show that it is real. Let

$$
Q = \Phi^2 - A^2 - 1 + 2i\Phi \quad\text{(D.2)}
$$

and denote the real part of Q as Q_R and the imaginary part as Q_I, i.e.,

$$
\begin{aligned}
Q_\mathrm{R} &= \Phi^2 - A^2 - 1 \quad\text{(D.3)}\\
Q_\mathrm{I} &= 2\Phi. \quad\text{(D.4)}
\end{aligned}
$$

We can also write (D.1) as

$$
\int_0^{2\pi} \frac{2(\Phi + A\sin\theta)}{1 + (\Phi + A\sin\theta)^2} d\theta = 2\mathrm{sgn}(\Phi)\left(Q^{-1/2} + \bar{Q}^{-1/2}\right) \quad\text{(D.5)}
$$

where $\bar{Q}$ is the complex conjugate of Q. For convenience, we also represent Q in the Euler form,

$$
Q = |Q|\, e^{j\omega}. \quad\text{(D.6)}
$$

Now we perform the simplification of the RHS of (D.5):

$$
\begin{aligned}
Q^{-1/2} + \bar{Q}^{-1/2} &= \left(|Q|\, e^{j\omega}\right)^{-1/2} + \left(|Q|\, e^{-j\omega}\right)^{-1/2}\\
&= |Q|^{-1/2}\left(e^{j\omega/2} + e^{-j\omega/2}\right)\\
&= 2|Q|^{-1/2}\cos\left(\frac{\omega}{2}\right). \quad\text{(D.7)}
\end{aligned}
$$

But from the half angle formula, $\cos\left(\dfrac{\omega}{2}\right) = \left(\dfrac{1 + \cos\omega}{2}\right)^{1/2}$, we have

$$\cos\left(\frac{\omega}{2}\right) = \left(\frac{1 + \frac{Q_R}{|Q|}}{2}\right)^{1/2}. \tag{D.8}$$

Thus (D.7) becomes

$$Q^{-1/2} + \bar{Q}^{-1/2} = \sqrt{2}\left(Q_R^2 + Q_I^2\right)^{-1/2}\left(Q_R + \left(Q_R^2 + Q_I^2\right)^{1/2}\right)^{1/2}. \tag{D.9}$$

Substituting (D.9) into (D.5) and using the definitions of Q_R and Q_I, we get

$$\int_0^{2\pi} \frac{2(\Phi + A\sin\theta)}{1 + (\Phi + A\sin\theta)^2}d\theta$$
$$= 2\sqrt{2}\pi\left(\left(\Phi^2 - A^2 - 1\right)^2 + 4\Phi^2\right)^{-1/2}$$
$$\times \left[\left(\left(\Phi^2 - A^2 - 1\right)^2 + 4\Phi^2\right)^{1/2} + \Phi^2 - A^2 - 1\right]^{1/2}. \tag{D.10}$$

Next, using Mathematica, we get

$$\int_0^{2\pi} \frac{2(\Phi + A\sin\theta)\sin\theta}{1 + (\Phi + A\sin\theta)^2}d\theta$$
$$= \frac{\pi}{A}\left[2 - (\Phi - i)\left(\Phi^2 - A^2 - 1 - 2i\Phi\right)^{-1/2}\right.$$
$$\left. - (\Phi + i)\left(\Phi^2 - A^2 - 1 + 2i\Phi\right)^{-1/2}\right]. \tag{D.11}$$

We show that this expression is real. With the definition of Q as before, the complex terms of (D.11) can be written as

$$(\Phi - i)\bar{Q}^{-1/2} + (\Phi + i)Q^{-1/2}$$
$$= \frac{1}{|Q|}\left[(\Phi - i)Q^{1/2} + (\Phi + i)\bar{Q}^{1/2}\right]$$
$$= \frac{1}{|Q|}\left[\left(\left(\Phi^2 - 1\right)Q_R + 2\Phi Q_I + i\left(\left(\Phi^2 - 1\right)Q_I - 2\Phi Q_R\right)\right)^{1/2}\right.$$
$$\left. + \left(\left(\Phi^2 - 1\right)Q_R + 2\Phi Q_I + i\left(\left(\Phi^2 - 1\right)Q_I - 2\Phi Q_R\right)\right)^{1/2}\right]. \tag{D.12}$$

Let

$$P = \left(\Phi^2 - 1\right)Q_R + 2\Phi Q_I + i\left(\left(\Phi^2 - 1\right)Q_I - 2\Phi Q_R\right)$$
$$= \left(\Phi^2 - 1\right)\left(\Phi^2 - A^2 - 1\right) + 4\Phi^2 + 2i\Phi A^2 \tag{D.13}$$

and denote the real part of P as P_R and the imaginary part as P_I , i.e.,

$$P_R = \left(\Phi^2 - 1\right)\left(\Phi^2 - A^2 - 1\right) \tag{D.14}$$

$$P_I = 2\Phi A^2 . \tag{D.15}$$

We can now write (D.11) as

$$\int_0^{2\pi} \frac{2(\Phi + A \sin\theta)\sin\theta}{1 + (\Phi + A\sin\theta)^2} d\theta = \frac{\pi}{A}\left(2 - \frac{1}{|Q|}\left(P^{1/2} + \bar{P}^{1/2}\right)\right) . \tag{D.16}$$

For convenience, we represent P in Euler form,

$$P = |P| e^{j\omega} . \tag{D.17}$$

Now we perform the simplification of the RHS of (D.16):

$$
\begin{aligned}
P^{1/2} + \bar{P}^{1/2} &= \left(|P| e^{j\omega}\right)^{1/2} + \left(|P| e^{-j\omega}\right)^{1/2} \\
&= 2|P|^{1/2} \cos\left(\frac{\omega}{2}\right) .
\end{aligned}
\tag{D.18}
$$

By the half angle formula, we get

$$\cos\left(\frac{\omega}{2}\right) = \left(\frac{1 + \frac{P_R}{|P|}}{2}\right)^{1/2} , \tag{D.19}$$

so (D.18) becomes

$$P^{1/2} + \bar{P}^{1/2} = \sqrt{2}\left(P_R + \left(P_R^2 + P_I^2\right)^{1/2}\right)^{1/2} . \tag{D.20}$$

Substituting (D.20) into (D.16) and using the definitions of P_R, P_I, and Q, we have

$$
\begin{aligned}
&\int_0^{2\pi} \frac{2(\Phi + A\sin\theta)\sin\theta}{1 + (\Phi + A\sin\theta)^2} d\theta \\
&= \frac{\pi}{A}\left\{2 - \sqrt{2}\left[\left(\left(\left(\Phi^2 - 1\right)\left(\Phi^2 - A^2 - 1\right) + 4\Phi^2\right)^2 + 4\Phi^2 A^4\right)^{1/2}\right.\right. \\
&\quad \left.\left. + \left(\Phi^2 - 1\right)\left(\Phi^2 - A^2 - 1\right) + 4\Phi^2\right]^{1/2}\right\} .
\end{aligned}
\tag{D.21}
$$

Appendix E

Derivation of the Critical Slopes

Define the RHS of (11.21) as $\mathcal{F}(R)$ and the RHS of (11.22) as $\mathcal{G}(R)$. Then $\mathcal{F}(R, \Phi)$ and $\mathcal{G}(R, \Phi)$ can be expressed as

$$
\begin{aligned}
\mathcal{F}(R, \Phi) &= (1 - \epsilon)\left(1 - \Phi^2 - R\right) \\
&\quad + \frac{2\epsilon}{3R}\left(1 - \frac{1}{\sqrt{2}k(R, \Phi)}\left(h(R, \Phi) + g(R, \Phi)\right)^{1/2}\right) \quad \text{(E.1)} \\
\mathcal{G}(R, \Phi) &= \Psi_{\text{C0}} + 1 + (1 - \epsilon)\left(\frac{3}{2}\Phi - \frac{1}{2}\Phi^3 - 3\Phi R\right) \\
&\quad + \frac{\sqrt{2}\,\mathrm{sgn}(\Phi)\,\epsilon}{k(R, \Phi)}\left(k(R, \Phi) + l(R, \Phi)\right)^{1/2}, \quad \text{(E.2)}
\end{aligned}
$$

where

$$
\begin{aligned}
l(R, \Phi) &= \Phi^2 - 4R - 1 & \text{(E.3)} \\
g(R, \Phi) &= \left(\Phi^2 - 1\right)l(R, \Phi) + 4\Phi^2 & \text{(E.4)} \\
h^2(R, \Phi) &= g^2(R, \Phi) + 64\Phi^2 R^2 & \text{(E.5)} \\
k^2(R, \Phi) &= l^2(R, \Phi) + 4\Phi^2. & \text{(E.6)}
\end{aligned}
$$

Differentiating $\mathcal{F}$ and $\mathcal{G}$ with respect to R and Φ, we get

$$
\begin{aligned}
\frac{\partial \mathcal{F}(R, \Phi)}{\partial R} &= (\epsilon - 1) - \frac{2\epsilon}{3R}\left(1 - \frac{1}{\sqrt{2}k(R, \Phi)}\left(h(R, \Phi) + g(R, \Phi)\right)^{1/2}\right) \\
&\quad - \frac{2\sqrt{2}\epsilon}{3Rh(R, \Phi)k^3(R, \Phi)\left(h(R, \Phi) + g(R, \Phi)\right)^{1/2}} \\
&\quad \times \left[2l(R, \Phi)h(R, \Phi)\left(h(R, \Phi) + g(R, \Phi)\right)\right. \\
&\quad \left. + k^2(R, \Phi)\left(\left(1 - \Phi^2\right)\left(h(R, \Phi) + g(R, \Phi)\right) + 16\Phi^2 R\right)\right] \quad \text{(E.7)}
\end{aligned}
$$

$$\frac{\partial \mathcal{F}(R, \Phi)}{\partial \Phi} = 2\Phi(\epsilon - 1) - \frac{2\sqrt{2}\Phi\epsilon}{h(R, \Phi)\, k^3(R, \Phi)\, (h(R, \Phi) + g(R, \Phi))^{1/2}}$$
$$\times \left[\left(\Phi^2 - 2R + 1\right)\left(g(R, \Phi) + h(R, \Phi)\, k^2(R, \Phi)\right)\right.$$
$$\left. - l(R, \Phi)\, h(R, \Phi)\, (h(R, \Phi) + g(R, \Phi)) + 16R^2\right] \qquad \text{(E.8)}$$

$$\frac{\partial \mathcal{G}(R, \Phi)}{\partial R} = -3(1 - \epsilon)\Phi - \frac{2\sqrt{2}\,\text{sgn}(\Phi)\,\epsilon}{k^3(R, \Phi)\, (k(R, \Phi) + l(R, \Phi))^{1/2}}$$
$$\times \left[k^2(R, \Phi) - l(R, \Phi)\, (k(R, \Phi) + 2l(R, \Phi))\right] \qquad \text{(E.9)}$$

$$\frac{\partial \mathcal{G}(R, \Phi)}{\partial \Phi} = (1 - \epsilon)\left(\frac{3}{2}\left(1 - \Phi^2\right) - 3R\right) + \frac{\sqrt{2}\Phi\,\text{sgn}(\Phi)\epsilon}{k^3(R, \Phi)\, (k(R, \Phi) + l(R, \Phi))^{1/2}}$$
$$\times \left(k^2(R, \Phi) - (l(R, \Phi) + 2)\, (k(R, \Phi) + 2l(R, \Phi))\right). \qquad \text{(E.10)}$$

Appendix F

Proof of Lemma 11.1

We start by rewriting (11.40) and (11.41) as

$$\mathcal{F}\left(A^2/4, \Phi\right) = \frac{1}{3\pi} \frac{2}{A} \int_0^{2\pi} \Psi_C \left(\Phi + A\sin\theta\right) \sin\theta d\theta \tag{F.1}$$

$$\mathcal{G}\left(A^2/4, \Phi\right) = \frac{1}{2\pi} \int_0^{2\pi} \Psi_C \left(\Phi + A\sin\theta\right) d\theta . \tag{F.2}$$

Then, the partial derivatives necessary for (11.63)–(11.66) are

$$\frac{\partial \mathcal{F}\left(A^2/4, \Phi\right)}{\partial \Phi} = \frac{1}{3\pi} \frac{2}{A} \int_0^{2\pi} \Psi'_C \left(\Phi + A\sin\theta\right) \sin\theta d\theta \tag{F.3}$$

$$\frac{\partial \mathcal{F}\left(A^2/4, \Phi\right)}{\partial R} = \frac{2}{A} \frac{1}{3\pi} \left[\frac{2}{A} \int_0^{2\pi} \Psi'_C \left(\Phi + A\sin\theta\right) \sin^2\theta d\theta \right.$$
$$\left. - \frac{2}{A^2} \int_0^{2\pi} \Psi_C \left(\Phi + A\sin\theta\right) \sin\theta d\theta \right] \tag{F.4}$$

$$\frac{\partial \mathcal{G}\left(A^2/4, \Phi\right)}{\partial \Phi} = \frac{1}{2\pi} \int_0^{2\pi} \Psi'_C \left(\Phi + A\sin\theta\right) d\theta \tag{F.5}$$

$$\frac{\partial \mathcal{G}\left(A^2/4, \Phi\right)}{\partial R} = \frac{1}{\pi A} \int_0^{2\pi} \Psi'_C \left(\Phi + A\sin\theta\right) \sin\theta d\theta . \tag{F.6}$$

When substituted into (11.63)–(11.66), all these expressions have to be evaluated at $\Phi = \Phi_{R+}(R) = \Phi_{R+}(A^2/4)$. The second term in (F.4) will become zero by definition of the function Φ_{R+}. Hence, we only need the function $\Psi'_C \left(\Phi_{R+}\left(A^2/4\right) + A\sin\theta\right)$. Since the Taylor expansion of Φ_{R+} is

$$\Phi_{R+}(A^2/4) = 1 + \Sigma_1(0)\frac{A^2}{4} + \frac{1}{2}\Sigma'_1(0)\left(\frac{A^2}{4}\right)^2 + O\left(A^6\right),$$

the function Ψ'_C can be approximated by

$$\Psi'_C \left(1 + A\sin\theta + S_1 \frac{A^2}{4} + \frac{1}{2}\Sigma'_1(0)\left(\frac{A^2}{4}\right)^2 + O\left(A^6\right)\right)$$

$$= \Psi''_C(1)\left[A\sin\theta + S_1 \frac{A^2}{4} + \frac{1}{2}\Sigma'_1(0)\left(\frac{A^2}{4}\right)^2 + O\left(A^6\right)\right]$$

$$+ \frac{1}{2}\Psi'''_C(1)\left[A\sin\theta + S_1 \frac{A^2}{4} + \frac{1}{2}\Sigma'_1(0)\left(\frac{A^2}{4}\right)^2 + O\left(A^6\right)\right]^2$$

$$+ \frac{1}{6}\Psi^{(4)}_C(1)\left[A\sin\theta + S_1 \frac{A^2}{4} + \frac{1}{2}\Sigma'_1(0)\left(\frac{A^2}{4}\right)^2 + O\left(A^6\right)\right]^3$$

$$+ \frac{1}{24}\Psi^{(5)}_C(1)\left[A\sin\theta + S_1 \frac{A^2}{4} + \frac{1}{2}\Sigma'_1(0)\left(\frac{A^2}{4}\right)^2 + O\left(A^6\right)\right]^4$$

$$+ O\left(\left[A\sin\theta + S_1 \frac{A^2}{4} + \frac{1}{2}\Sigma'_1(0)\left(\frac{A^2}{4}\right)^2 + O\left(A^6\right)\right]^5\right)$$

$$= \Psi''_C(1)\sin\theta A + \left(\frac{1}{4}\Psi''_C(1)S_1 + \frac{1}{2}\Psi'''_C(1)\sin^2\theta\right)A^2$$

$$+ \left(\frac{1}{4}\Psi'''_C(1)S_1\sin\theta + \frac{1}{6}\Psi^{(4)}_C(1)\sin^3\theta\right)A^3$$

$$+ \left(\frac{1}{32}\Psi''_C(1)\Sigma'_1(0) + \frac{1}{32}\Psi'''_C(1)S_1^2\right.$$

$$\left. + \frac{1}{12}\Psi^{(4)}_C(1)S_1\sin^2\theta + \frac{1}{24}\Psi^{(5)}_C(1)\sin^4\theta\right)A^4 + O\left(A^5\right). \qquad \text{(F.7)}$$

By substituting (F.7) into (F.3)–(F.6), we get

$$\left.\frac{\partial\mathcal{F}(R,\Phi)}{\partial\Phi}\right|_{\Phi=\Phi_{R+}(R)} = \frac{2}{3}\left[\Psi''_C(1) + \left(\Psi'''_C(1)S_1 + \frac{1}{2}\Psi^{(4)}_C(1)\right)R\right]$$

$$+ O\left(R^2\right) \qquad \text{(F.8)}$$

$$\left.\frac{\partial\mathcal{F}(R,\Phi)}{\partial R}\right|_{\Phi=\Phi_{R+}(R)} = \frac{1}{3}\left[\Psi''_C(1)S_1 + \frac{3}{2}\Psi'''_C(1) + \left(\frac{1}{2}\Psi''_C(1)\Sigma'_1(0)\right.\right.$$

$$\left.\left. + \frac{1}{2}\Psi'''_C(1)S_1^2 + \Psi^{(4)}_C(1)S_1 + \frac{5}{12}\Psi^{(5)}_C(1)\right)R\right]$$

$$+ O\left(R^2\right) \qquad \text{(F.9)}$$

$$\left.\frac{\partial\mathcal{G}(R,\Phi)}{\partial\Phi}\right|_{\Phi=\Phi_{R+}(R)} = \left(\Psi''_C(1)S_1 + \Psi'''_C(1)\right)R + O\left(R^2\right) \qquad \text{(F.10)}$$

$$\frac{\partial \mathcal{G}(R,\Phi)}{\partial R}\bigg|_{\Phi=\Phi_{R+}(R)} = \Psi_C''(1) + \left(\Psi_C'''(1)S_1 + \frac{1}{2}\Psi_C^{(4)}(1)\right)R$$
$$+ O\left(R^2\right). \tag{F.11}$$

By substituting (F.8) into (11.63), we get

$$a_1(R) = -\frac{2\sigma}{3}\Psi_C''(1)R + O\left(R^2\right). \tag{F.12}$$

By substituting (F.10) into (11.64), we get

$$a_2(R) = -\left(\Psi_C''(1)S_1 + \Psi_C'''(1)\right)R + O\left(R^2\right). \tag{F.13}$$

By substituting (F.8) and (F.9) into (11.65), we get

$$\Sigma_1(R)$$
$$= -\left(\frac{1}{2}S_1 + \frac{3}{4}\frac{\Psi_C'''(1)}{\Psi_C''(1)}\right)$$
$$+ \frac{1}{2\Psi_C''(1)}\left[\left(S_1 + \frac{3}{2}\frac{\Psi_C'''(1)}{\Psi_C''(1)}\right)\left(\Psi_C'''(1)S_1 + \frac{1}{2}\Psi_C^{(4)}(1)\right)\right.$$
$$\left. - \left(\frac{1}{2}\Psi_C''(1)\Sigma_1'(0) + \frac{1}{2}\Psi_C'''(1)S_1^2 + \Psi_C^{(4)}(1)S_1 + \frac{5}{12}\Psi_C^{(5)}(1)\right)\right]R$$
$$+ O\left(R^2\right). \tag{F.14}$$

Since $\Sigma_1(0) = S_1 = -\left(\frac{1}{2}S_1 + \frac{3}{4}\Psi_C'''(1)/\Psi_C''(1)\right)$, we get

$$S_1 = -\frac{1}{2}\frac{\Psi_C'''(1)}{\Psi_C''(1)}. \tag{F.15}$$

In a similar fashion, we determine $\Sigma_1'(0)$ and get

$$\Sigma_1(R) = -\frac{1}{2}\frac{\Psi_C'''(1)}{\Psi_C''(1)} + \left[\frac{5}{4}S_1^3 - \frac{1}{4S_2}\left(3\Psi_C^{(4)}(1)S_1 + \frac{5}{12}\Psi_C^{(5)}(1)\right)R\right]$$
$$+ O\left(R^2\right). \tag{F.16}$$

Finally, by substituting (F.11), (F.13), and (F.16) into (11.66), we get

$$\Sigma_2(R) = \Psi_C''(1) + \left(\frac{3}{2}\Psi_C'''(1)S_1 + \frac{1}{2}\Psi_C^{(4)}(1)\right)R + O\left(R^2\right)$$
$$\triangleq S_2 + \left(-3S_1^2 S_2 + \frac{1}{2}\Psi_C^{(4)}(1)\right)R + O\left(R^2\right). \tag{F.17}$$

We conclude the proof by noting that (11.71) and (11.72) are obtained by substituting $S_1 = -\frac{1}{2}\Psi_C'''(1)/\Psi_C''(1)$ and $S_2 = \Psi_C''(1)$ into (F.12) and (F.13).

Bibliography

[1] R. A. Adomaitis, and E. H. Abed, "Local nonlinear control of stall inception in axial flow compressors," AIAA Paper #93-2230, *29th Joint Propulsion Conference*, Monterey, CA, June 1993.

[2] A. M. Annaswamy and A. F. Ghoniem, "Active control in combustion systems," *IEEE Control Systems Magazine*, vol. 15, pp. 49–63, Dec. 1995.

[3] A. M. Annaswamy and A. F. Ghoniem, "Active control of combustion instability: theory and practice," *IEEE Control Systems Magazine*, vol. 22, pp. 37–54, Dec. 2002.

[4] M. Arcak and P. V. Kokotović, "Nonlinear observers: a circle criterion design and robustness analysis", *Automatica*, vol. 37, pp. 1923–1930, 2001.

[5] K. B. Ariyur and M. Krstić, "Analysis and design of multivariable extremum seeking", *2002 American Control Conference*, Anchorage, AK, pp. 2903–2908, May 2002.

[6] K. B. Ariyur and M. Krstić, "Slope seeking and application to compressor instability control", *Proceedings of the IEEE Conference on Decision and Control*, Las Vegas, NV, pp. 3690–3697, Dec. 2002.

[7] K. J. Astrom and B. Wittenmark, *Adaptive Control*, 2nd ed., Addison-Wesley, Reading, MA, 1995.

[8] O. O. Badmus, S. Chowdhury, K. M. Eveker, C. N. Nett, and C. J. Rivera, "A simplified approach for control of rotating stall, Parts I and II", *Proceedings of the 29th Joint Propulsion Conference*, AIAA papers 93-2229 & 93–2234, Monterey CA, June 1993.

[9] O. O. Badmus, C. N. Nett, and F. J. Schork, "An integrated, full-range surge control/rotating stall avoidance compressor control system," *Proceedings of the American Control Conference*, pp. 3173–3180, Boston, MA, 1991.

[10] E.-W. Bai, L.-C. Fu, and S. S. Sastry, "Averaging analysis for discrete time and sampled data adaptive systems," *IEEE Trans. Circuits and Systems*, vol. 35, pp. 137-148, 1988.

[11] A. Banaszuk, K. B. Ariyur, M. Krstić and C. A. Jacobson, "An adaptive algorithm for control of combustion instability," *Automatica*, to appear.

[12] A. Banaszuk, C. A. Jacobson, A. I. Khibnik and P. G. Mehta, "Linear and nonlinear analysis of controlled combustion processes. Part I: Linear analysis, Part II: Nonlinear analysis," *Proceedings of the IEEE Conference on Control Applications*, Kohala-Coast, HI, pp. 199–212, Aug. 1999.

[13] A. Banaszuk and A. J. Krener, "Design of controllers for MG3 compressor models with general characteristics using graph backstepping," *Automatica*, vol. 35, pp. 1343–1368, 1999.

[14] A. Banaszuk, Y. Zhang and C. A. Jacobson , " Active Control of Combustion Instabilities in Gas Turbine Engines for Low Emissions. Part II: Adaptive Algorithm Development, Demonstration, and Performance Limitations ," *Proceedings of the NATO AVT/RTO Symposium*, Braunschewig, Germany, 2000.

[15] A. Banaszuk and Y. Zhang and C. A. Jacobson , "Adaptive control of combustion instability using extremum-seeking," *Proceedings of the American Control Conference*, pp. 416–422, Chicago, IL, June 2000.

[16] R. N. Banavar, D. F. Chichka, J. L. Speyer, "Convergence and synthesis issues in extremum seeking control," *Proceedings of the American Control Conference*, Chicago, IL, pp. 438–443, June 2000.

[17] G. Bastin and D. Dochain, *On-line Estimation and Adaptive Control of Bioreactors*, Elsevier Science Publications, New York, 1990.

[18] G. Bastin and J. F. Van Impe, "Nonlinear and adaptive control in biotechnology: a tutorial," *European Journal of Control*, pp. 37–53, vol. 1, 1995.

[19] R. L. Behnken, R. D'Andrea, and R. M. Murray, "Control of rotating stall in a low-speed axial flow compressor using pulsed air injection: modeling, simulations, and experimental validation," *Proceedings of the 34th IEEE Conference on Decision and Control*, pp. 3056–3061, 1995.

[20] P. Binetti, K. B. Ariyur, M. Krstić and F. Bernelli, "Formation flight optimization using extremum seeking feedback", *AIAA Journal of Guidance, Control, and Dynamics*, pp. 132–142, vol. 26, January-February 2003.

[21] P. F. Blackman, "Extremum-Seeking Regulators," in J. H. Westcott, Ed., *An Exposition of Adaptive Control*, The Macmillan Company, New York, NY, 1962.

[22] W. B. Blake and D. Multhopp, "Design performance and modeling considerations for closed formation flight", *AIAA Atmospheric Flight Mechanics Conference*, Boston, MA, AIAA-98-4343, Aug. 1998.

[23] W. B. Blake, "An aerodynamic model for simulation of closed formation flight", *AIAA Modeling and Simulation Technologies Conference*, Denver, CO, AIAA-2000-4034, Aug. 2000

[24] J. D. Bosković and K. S. Narendra, "Comparison of linear, nonlinear and neural-network-based adaptive controllers for a class of fed-batch fermentation processes," *Automatica*, pp. 817–840, vol. 31, 1995.

[25] L. Chen, G. Bastin, and V. V. Breusegem, "A case study of adaptive nonlinear regulation of fed-batch biological reactors," *Automatica*, pp. 55–65, vol. 31, 1995.

[26] D. F. Chichka and J. Speyer, "Solar-powered, formation-enhanced aerial vehicle systems for sustained endurance", *Proceedings of the American Control Conference*, pp. 684–688, Philadelphia, PA, June 1998.

[27] D. F. Chichka, J. Speyer and C. G. Park, "Peak-seeking control with application to formation flight", *Proceedings of the 38th IEEE Conference on Decision and Control*, pp. 2463–2470, Phoenix, AZ, December 1999.

[28] Y. K. Chin *et al*, "Sliding model ABS wheel slip control," *Proceedings of the American Control Conference*, pp. 1–6, 1992.

[29] P. I. Chinaev, Ed., *Self-Tuning Systems Handbook*, in Russian, Naukova Dumka, Kiev, 1969.

[30] J. Y. Choi, M. Krstić, K. B. Ariyur and J. S. Lee, "Extremum seeking control for discrete-time systems," *IEEE Transactions on Automatic Control*, vol. 47, pp. 318–323, 2002.

[31] J. M. Cohen, N. M. Rey, C. A. Jacobson and T. J. Anderson, "Active control of combustion instability in a liquid-fueled low-NO_x combustor," *ASME/IGTI Gas turbine Expo and Congress*, Stockholm, Sweden, June 1998.

[32] C. J. Cutts and J. R. Speakman, "Energy savings in formation flight of pink-footed geese", *Journal of Experimental Biology*, Vol. 189, January 1994, pp. 251–261.

[33] R. D'Andrea, R. L. Behnken, and R. M. Murray, "Active control of rotating stall using pulsed air injection: A parametric study on a low-speed, axial slow compressor," In *Proceedings of SPIE*, volume 2494, pages 152–165, Orlando Florida, 1995.

[34] G. D'Ans and P. Kokotović, "Optimal control of bacterial growth," *Automatica*, pp. 729–736, 1972.

[35] S. Drakunov, U. Ozguner, P. Dix, and B. Ashrafi, "ABS control using optimum search via sliding modes," *IEEE Transactions on Control Systems Technology*, vol. 3, pp. 79–85, 1995.

[36] C. S. Drapper and Y. T. Li, "Principles of optimalizing control systems and an application to the internal combustion engine," *ASME*, vol. 160, pp. 1–16, 1951, also in: R. Oldenburger (Ed.), Optimal and Self-Optimizing Control, MIT Press, Boston, MA, 1966.

[37] E. Elong, M. Krstić and K. B. Ariyur, "A case study of performance improvement in extremum seeking control," *Proceedings of the American Control Conference*, Chicago, IL, pp. 428–432, June 2000.

[38] K. M. Eveker, D. L. Gysling, C. N. Nett, and O. P. Sharma, "Integrated control of rotating stall and surge in aeroengines," *1995 SPIE Conference on Sensing, Actuation, and Control in Aeropropulsion*, Orlando, April 1995.

[39] A. A. Feldbaum, *Computers in Automatic Control Systems*, in Russian, Fizmatgiz, Moscow, 1959.

[40] D. Fontaine, S. Liao, J. Paduano and P. Kokotović, "Linear vs. nonlinear control of an axial flow compressor," *Proceedings of the IEEE Conference on Control Applications*, Kohala Coast, HI, pp. 921–926, 1999.

[41] G. F. Franklin, J. D. Powell and A. Emami-Naeini, "Integral Control and Robust Tracking", *Feedback Control of Dynamic Systems*, 3rd ed., Addison-Wesley, 1995, pp. 551–560.

[42] A. L. Frey, W. B. Deem, and R. J. Altpeter, "Stability and optimal gain in extremum-seeking adaptive control of a gas furnace," *Proceedings of the Third IFAC World Congress*, London, 48A, 1966.

[43] L. J. Garodz, "Measurement of the wake characteristics of the Boeing 747, Lockheed C-5A, and other aircraft", NASA, Project 177-621-03X (Special Task No. 1), NAFEC, Atlantic City, NJ, Apr. 1970.

[44] R. Genesio, M. Basso, and A. Tesi, Analysis and Synthesis of Limit Cycle Bifurcations in Feedback Systems, *Proceedings of the 34th CDC*, New Orleans, LA, December, 1995.

[45] F. Giulietti, L. Pollini, and M. Innocenti, "Autonomous formation flight", *IEEE Controls Systems Magazine*, Vol. 20, No. 6, 2000, pp. 34–44.

[46] G. C. Goodwin and K. S. Sin, *Adaptive Filtering Prediction and Control*, Englewood Cliffs, NJ: Prentice-Hall, 1984.

[47] E. M. Greitzer, "Surge and rotating stall in axial flow compressors—Part I: Theoretical compression system model," *Journal of Engineering for Power*, pp. 190–198, 1976.

[48] G. Gu, A. Sparks, and S. Banda, "Bifurcation based nonlinear feedback control for rotating stall in axial compressors," *International Journal of Control*, vol. 68, pp.1241-1257, 1997.

[49] J. N. Hallock and W. R. Eberle, "Aircraft wake vortices: a state of the art review of the United States R&D program", DOT, Transportation Systems Center, DOT-TSC-FAA-77-4, Cambridge, MA, Feb. 1977.

[50] J. N. Hallock, "Aircraft wake vortices: an assessment of the current situation", DOT, Research and Special Programs Administration, DOT-TSC-FAA-90-6, Cambridge, MA, Jan. 1991.

[51] I. Haskara, U. Ozguner and J. Winkelman, "Extremum control for operating point determination and set point optimization via sliding modes," *Journal of Dynamic Systems, Measurement, and Control*, vol. 122, pp. 719–724, 2000.

[52] J. P. Hathout, A. M. Annaswamy, M. Fleifil and A. F. Ghoniem, "A model-based active control design for thermoacoustic instability", *Combustion Sci. and Tech.*, vol. 132, pp. 99–105, 1998.

[53] R. K. Heffley and W. F. Jewell, "Aircraft handling qualities data", Systems Technology Inc., CR-2144, Washington, DC, Dec. 1972.

[54] D. Herbert, R. Elsworth, and R. C. Telling, "The continuous culture of bacteria; a theoretical and experimental study," *Journal of General Microbiology*, pp. 601–622, 1956.

[55] W. Hui, B. A. Bamieh and G. H. Miley, "Robust Burn Control of a Fusion Reactor by Modulation of the Refueling Rate", *Fusion Technology*, vol. 25, pp. 318–25, May 1994.

[56] D. Hummel, "The use of aircraft wakes to achieve power reduction in formation flight", *Proceedings of the Fluid Dynamics Panel Symposium*, AGARD, Trondheim, Norway, May 1996, pp. 1777–1794.

[57] P. A. Ioannou and J. Sun, *Stable and Robust Adaptive Control*, Englewood Cliffs, NJ: Prentice-Hall, 1995.

[58] A. Isidori, *Nonlinear Control Systems II*, Springer-Verlag, London, UK, 2000.

[59] O. L. R. Jacobs and G. C. Shering, "Design of a single-input sinusoidal-perturbation extremum-control system," *Proceedings of the IEE*, vol. 115, pp. 212–217, 1968.

[60] M. Janković, "Stability analysis and control of compressors with noncubic characteristic," PRET Working Paper B95-5-24, Center for Control Engineering and Computation, University of California at Santa Barbara, 1995.

[61] C. E. Johnson, Y. Neumeier, E. Lubarsky, Y. J. Lee, M. Neumaier, and B.T. Zinn, "Suppression of combustion instabilities in a liquid fuel combustor using a fast adaptive algorithm," *AIAA Paper AIAA-2000-0476*, 38th Aerospace Sciences Meeting & Exhibit, Reno, January 2000.

[62] S. W. Kandebo, "C-5 Reengining Should Boost Performance", *Aviation Week and Space Technology*, Vol. 153, No. 7, 2000, pp. 26–27.

[63] V. V. Kazakevich, "Extremum control of objects with inertia and of unstable objects," *Soviet Physics, Dokl. 5*, pp. 658–661, 1960.

[64] H. K. Khalil, *Nonlinear Systems*, 2nd edition, Prentice-Hall, Upper Saddle River, NJ, 1996.

[65] S. K. Korovin and V. I. Utkin,"Using sliding modes in static optimization and nonlinear programming," *Automatica*, vol. 10, pp. 525–532, 1974.

[66] A. A. Krasovskii, *Dynamics of Continuous Self-Tuning Systems*, in Russian, Fizmatgiz, Moscow, 1963.

[67] M. Krstić, "Performance improvement and limitations in extremum seeking control," *Systems & Control Letters*, vol. 39, pp. 313–326, 2000.

[68] M. Krstić, D. Fontaine, P. V. Kokotović and J. D. Paduano,"Useful nonlinearities and global bifurcation control of jet engine surge and stall," *IEEE Transactions on Automatic Control*, vol. 43, pp. 1739–1745, 1998.

[69] M. Krstić, I. Kanellakopoulos, and P. V. Kokotović, *Nonlinear and Adaptive Control Design*, New York: Wiley, 1995.

[70] M. Krstić and H. H. Wang, "Stability of extremum seeking feedback for general nonlinear dynamic systems," *Automatica*, vol. 36, pp. 595–601, 2000.

[71] W. Lang, T. Poinsot and S. Candel, "Active control of combustion instability", *Combustion and Flame*, vol. 70, pp. 281–289, 1987.

[72] B. La Scala, "Approaches to Frequency Tracking and Vibration Control," *Ph.D. Thesis*, Dept. of Systems Engineering, The Australian National University, December 1994.

[73] M. Leblanc, "Sur l'electrification des chemins de fer au moyen de courants alternatifs de frequence elevee," *Revue Generale de l'Electricite*, 1922.

[74] Z.-H. Li, *Optimal Lyapunov Design of Robust and Adaptive Nonlinear Controllers*, PhD thesis, University of Maryland, College Park, 1997.

[75] D.-C. Liaw and E. H. Abed, "Active control of compressor stall inception: a bifurcation-theoretic approach," *Automatica*, vol. 32, pp. 109–115, 1996.

[76] G. R. Ludwig and J. P. Nenni, "Tests of an improved rotating stall control system on a J-85 turbojet engine," *ASME* Paper 80-GT-17, 1980.

[77] C. A. Mansoux, J. D. Setiawan, D. L. Gysling, and J. D. Paduano, "Distributed nonlinear modeling and stability analysis of axial compressor stall and surge," *1994 American Control Conference*, Baltimore, July 1994.

[78] I. Mareels and J. W. Polderman, *Adaptive Systems: An Introduction*, Springer Verlag, Berlin, June 1996.

[79] B. Maskew, "Formation Flying Benefits Based on Vortex Lattice Calculations", Analytical Methods Inc., CR-151974, Bellevue, WA, 1977

[80] F. E. McCaughan, "Bifurcation analysis of axial flow compressor stability," *SIAM Journal of Applied Mathematics*, vol. 20, pp. 1232–1253, 1990.

[81] S. M. Meerkov, "Asymptotic Methods for Investigating Quasistationary States in Continuous Systems of Automatic Optimization," *Automation and Remote Control*, no. 11, pp. 1726–1743, 1967.

[82] S. M. Meerkov, "Asymptotic Methods for Investigating a Class of Forced States in Extremal Systems," *Automation and Remote Control*, no. 12, pp. 1916–1920, 1967.

[83] S. M. Meerkov, "Asymptotic Methods for Investigating Stability of Continuous Systems of Automatic Optimization Subjected to Disturbance Action," (in Russian) *Avtomatika i Telemekhanika*, no. 12, pp. 14–24, 1968.

[84] F. K. Moore, and E. M. Greitzer, "A theory of post-stall transients in axial compression systems—Part I: Development of equations," *Journal of Engineering for Gas Turbines and Power*, vol. 108, pp. 68–76, 1986.

[85] I. S. Morosanov, "Method of extremum control," *Automatic & Remote Control*, vol. 18, pp. 1077–1092, 1957.

[86] S. Murugappan, E. J. Gutmark, and S. Acharya, "Application of extremum-seeking controller to suppression of combustion instabilities in spray combustion," *AIAA Paper AIAA-2000-1025*, 38th Aerospace Sciences Meeting & Exhibit, Reno, January 2000.

[87] J. H. Myatt and W. B. Blake, "Aerodynamic database issues for modeling close formation flight", *AIAA Modeling and Simulation Technologies Conference and Exhibit*, Portland, OR, AIAA-99-4194, Aug. 1999.

[88] M. R. Myers, D. L. Gysling, and K. M. Eveker, "Benchmark for control design: Moore-Greitzer model," PRET Working Paper, UTRC95-9-18.

[89] K. S. Narendra and A. M. Annaswamy, *Stable Adaptive Systems*, Prentice Hall, Englewood Cliffs, NJ, 1989.

[90] I. I. Ostrovskii, "Extremum regulation," *Automatic & Remote Control*, vol. 18, pp. 900–907, 1957.

[91] M. Pachter, J. J. D'Azzo and A. W. Proud, "Tight formation flight control", *AIAA Journal of Guidance, Control, and Dynamics*, Vol. 24, No. 2, 2001, pp. 246–254.

[92] A. A. Pervozvanskii, "Continuous extremum control system in the presence of random noise," *Automation and Remote Control*, vol. 21, pp. 673–677, 1960.

[93] K. S. Peterson, A. G. Stefanopoulou and Y. Wang, "Extremum seeking control for soft landing of an electromechanical valve actuator," Submitted to *Automatica*, January 2003.

[94] D. Popović, M. Janković, S. Magner and A. Teel, "Extremum seeking methods for optimization of variable cam timing engine operation," To appear in *Proceedings of the 2003 American Control Conference*, Denver, CO, June 2003.

[95] J. Ramsey, "C-5 modernization-step by step", *Avionics Magazine*, Vol. 24, No. 8, 2000, pp. 30–34.

[96] M. A. Rotea, "Analysis of multivariable extremum seeking algorithms," *Proceedings of the American Control Conference*, Chicago, IL, pp. 433–437, June 2000.

[97] W. J. Rugh, *Linear System Theory*, Prentice Hall, Upper Saddle River, NJ, 1996.

[98] S. Sastry and M. Bodson, *Adaptive Control: Stability, Convergence, and Robustness*, Englewood Cliffs, NJ:Prentice-Hall, 1989.

[99] J. Scheiman, J. L. Megrail and J. P. Shivers, "Exploratory Investigation of Factors Affecting the Wing Tip Vortex", NASA Langley Research Center, TM X-2516, Hampton, VA, April 1972.

[100] G. Schneider, K. B. Ariyur and M. Krstić, "Tuning of a combustion controller by extremum seeking: A simulation study," *Conference on Decision and Control*, Sydney, Australia, pp. 5219–5223, Dec. 2000.

[101] C. Schumacher and S. Singh, "Nonlinear control of multiple UAVs in closed coupled formation flight", *AIAA Guidance, Navigation, and Control Conference*, Denver, CO, AIAA-2000-4373, Aug. 2000.

[102] R. Sepulchre and P. V. Kokotović, "Shape signifiers for control of a low-order compressor model," *IEEE Transactions on Automatic Control*, vol. 43, pp. 1643–1648, 1998.

[103] J. R. Seume, N. Vortmeyer, W. Krause, J. Hermann, C.-C. Hantschk, P. Zangl, S. Gleis, D. Vortmeyer, and A. Orthmann,"Application of Active Combustion Instability Control to a Heavy Duty Gas Turbine, ", *Proceedings of ASME Asia '97 Congress and Exhibition*, Singapore, October 1997, ASME Paper 97-AA-119.

[104] M. E. Sezer and D. D. Siljak, "Decentralized control," in *Control Handbook*, W. S. Levine Ed., pp. 779–793, CRC Press, 1996.

[105] J. Sternby, " Extremum control systems: An area for adaptive control?," Preprints of the *Joint American Control Conference*, San Francisco, CA, 1980, WA2-A.

[106] H. S. Tan and M. Tomizuka, "An adaptive sliding model vehicle traction controller design," *Proceedings of American Control Conference*, 1035–1058, 1989.

[107] A. R. Teel and D. Popović, "Solving smooth and nonsmooth multivariable extremum seeking problems by the methods of nonlinear programming," *Proceedings of the American Control conference*, Arlington, VA, pp. 2394–2399, June 2001.

[108] J. Thibault, V. V. Breusegem, and A. Chéruy, "On-line prediction of fermentation variables using neural networks," *Biotechnology and Bioengineering*, pp. 1041–48, 1990.

[109] H. S. Tsien, *Engineering Cybernetics*, New York, NY: McGraw-Hill, 1954.

[110] I. Tunay, "Antiskid control for aircraft via extremum seeking," *Proceedings of the American Control Conference*, Arlington, VA, pp. 665–670, June 2001.

[111] J. F. Van Impe and G. Bastin, "Optimal adaptive control of fed-batch fermentation processes with multiple substrates," *Second IEEE Conference on Control Applications*, pp. 469–474, Vancouver, BC, Canada, 1993.

[112] G. Vasu, "Experiments with optimizing controls applied to rapid control of engine pressures with high amplitude noise signals," *Transactions of the ASME*, pp. 481–488, 1957.

[113] G. C. Walsh, "On the application of multi-parameter extremum seeking control," *Proceedings of the American Control Conference*, Chicago, IL, pp. 411–415, June 2000.

[114] H.-H. Wang and M. Krstić, "Extremum seeking for limit cycle minimization," *IEEE Transactions on Automatic Control*, vol. 45, pp. 2432–2436, Dec. 2000..

[115] H.-H. Wang, M. Krstić, and G. Bastin, "Optimizing bioreactors by extremum seeking," *International Journal of Adaptive Control and Signal Processing*, vol. 13, pp. 651–659, 1999.

[116] H. -H. Wang, M. Krstić and M. Larsen, "Control of deep hysteresis aeroengine compressors", *Journal of Dynamic Systems, Measurement, and Control*, vol. 122, pp. 140–152, 2000.

[117] H. H. Wang, S. Yeung and M. Krstić, "Experimental application of extremum seeking on an axial-flow compressor," *IEEE Trans. on Control Systems Technology*, vol. 8, pp. 300–309, 2000.

[118] Y. Wang and R. M. Murray, "A geometric perspective on bifurcation control," *Proceedings of the 39th IEEE Conference on Decision and Control*, Sydney, Australia, pp. 1613–1618, Dec. 2000.

[119] D. J. Wilde, *Optimum Seeking Methods*, Prentice Hall, 1964.

[120] W. Williamson, J. Min, J. L. Speyer and J. Farrell, "A comparison of state space, range space, and carrier phase differential GPS/INS relative navigation", *Proceedings of the 2000 American Control Conference*, Chicago, IL, June 2000, pp. 2932–2938.

[121] Y. Y. Yang and D. A. Linkens, "Adaptive neural-network-based approach for the control of continuously stirred tank reactor," *IEE Proceedings-Control Theory and Applications*, vol. 141, p. 341-9, 1994.

[122] Y. Zhang, "Stability and performance tradeoff with discrete time triangular search minimum seeking," *Proceedings of the American Control Conference*, Chicago, IL, pp. 423-427, June 2000.

[123] K. Zhou, J. C. Doyle and K. Glover, *Robust and Optimal Control*, Prentice-Hall, Upper Saddle River, NJ, 1995.

Index